STUDENT ORGANIZER

GEX, INCORPORATED

BEGINNING ALGEBRA
SIXTH EDITION

Elayn Martin-Gay

University of New Orleans

PEARSON

Boston   Columbus   Indianapolis   New York   San Francisco   Upper Saddle River
Amsterdam   Cape Town   Dubai   London   Madrid   Milan   Munich   Paris   Montreal   Toronto
Delhi   Mexico City   Sao Paulo   Sydney   Hong Kong   Seoul   Singapore   Taipei   Tokyo

ISBN-13: 978-0-321-78521-3
ISBN-10: 0-321-78521-5

 4 5 6 EBM 16 15 14 13

www.pearsonhighered.com

PEARSON

Table of Contents

Organizer Overview (for Instructors)

Greetings and thank you so much for using my Organizer. Upon completion of using this tool, I welcome any comments that will help me do a better job.

Let me describe to you what I see as a major problem in mathematics and the solution I am trying to provide:

Problem: Many of our students come to mathematics courses lacking not only the understanding of mathematical concepts and skills needed for high school mathematics, but also the general organizational/study tools needed for success. These helpful tools-for-success include general note-taking in mathematics, notebook organization, and basic study methods that are particularly successful in mathematics. You have probably noticed that many of our students, even those who participate successfully in the classroom, have trouble starting and documenting their assigned homework, reading their text, and maybe most importantly, keeping organized coursework.

Solution: This supplement is to be actively used with Martin-Gay's *Beginning Algebra*, 6th ed. If used properly, it will enable each student to grow in the skills listed above and to be successful and fully organized in their current course.

The Student Organizer contains:

> Organizer Overview for Students (printed front and back)

> For each section of the text, there are two 3-hole punched pages (printed front and back), with each page divided into sections focusing on needed organization.

> The final page for each chapter contains help for preparing for a chapter test.

> **Each student does need a three-ring binder (notebook), preferably with pockets.**

How does the Organizer Work?

Please read the Organizer Overview (for Students) for detailed instructions. You will see how this Organizer helps a student start reading this text, and start using some available supplements.

As an instructor, you can easily customize this supplement. Your course syllabus, etc., can be placed at the front of this notebook. If you give quizzes, students can place each within the appropriate section or in a notebook section solely for that purpose. Tests can be placed in the notebook pockets or in another designated section of the notebook.

Thank you, again, and best wishes for your Beginning Algebra course,
Elayn (Martin-Gay)

Organizer Overview (for Students)

Greetings and thank you for using my Organizer.

This supplement is to be placed in a 3-ring notebook (preferably with pockets) to be used along with Martin-Gay's *Beginning Algebra*, 6th ed. If used as directed, this Organizer will help you become better organized, use your text more efficiently, use the Lecture Videos that are provided for you, and ultimately, increase your study skills for not only this course, but for other classes you are taking now and in the future.

How to Use this Supplement:

First, make sure you have the proper supplies.

Tools needed: Text, Video Lecture Series, writing instrument (pencil or pen), and this Organizer in a three-holed notebook.

For each section of each chapter, there are two 3-hole punched pages printed front and back, with each page divided into sections focusing on needed organization. These sections are structured as noted below.

Page 1 (front and back) contains:

> **Before Class**: Read and follow the given directions. Place a checkmark within the small open square before each set of directions as you complete each set.

> **During Class:** This section has to do with writing notes that will be helpful to you later as you start homework or study for a quiz or test. Notice that the remainder of this page along with the back of the page is divided into two columns.

>> **Class Notes/Examples**: In this column, write down any examples (line-by-line) demonstrated by your instructor, seen as an example in MyMathLab, or in the Lecture Videos.

>> **Your Notes**: This smaller column to the right is for your personal notes; for example, for you to write down things you don't want to forget.

> **After this page**, please insert any additional paper of your own to write any further notes.

Organizer Overview (for Students)

Page 2 (front and back) contains:

> **Practice:** Read and follow the given directions. There are numbered examples and exercises for you to read and/or complete. For each of these, the answers and/or references are at the end of the section for your review. The following are types of exercises/examples:
>
>> **Review this example**: This example is shown worked and completed and the answer is circled. Read this example and make sure you understand the solution.
>>
>> **Your turn**: This exercise is for you to work and circle when completed. Make sure you check the answer at the end of this section. If correct, move to the next example/exercise. If incorrect, use the reference by the answer to view detailed steps of the solution.
>>
>> **Complete this example**: This example is partially complete. Read the completed part and fill in the blank(s). (Follow the same steps as above to check and correct your work.)
>
> **After this page**, please insert your paper containing your *written homework* * assigned by your instructor.

---

* *Written Homework*: Attempt all exercises asked of you. All odd answers are in the e-book, so make sure you check the answers to these exercises. If an exercise answer is incorrect, try to correct it on your own. If you are unable to correct an exercise, place a mark by the exercise number (such as a question mark, "?") so that you will know to ask your instructor about it. If there is an exercise that you want to make sure you study again before a test, place a mark (maybe an "!" mark) that you will recognize later.

Follow these directions as closely as possible. I know this may be difficult in the beginning, but trust that this Organizer can help you with your mathematics course.

Best wishes to you in your Beginning Algebra course,
Elayn Martin-Gay

Section 1.2 Symbols and Sets of Numbers

Before Class:

☐   Read the objectives on page 7.

☐   Read the **Helpful Hint** boxes on pages 7, 11, and 13.

☐   Complete the exercises:

   1.   The symbols $\neq, <,$ and $>$ are called _____ .

   2.   Look at the Common Sets of Numbers diagram on page 11. From the diagram, all integers are also elements of what two other sets?

   3.   Since $|a|$ is the distance between a and 0 on a number line, $|a|$ is never _____ .

During Class:

☐   **Write your class notes.** Neatly write down **all** examples shown as well as key terms or phrases with definitions. If not applicable or if you were absent, watch the Lecture Series (DVD) for this section and do the same (write down the examples shown as well as key terms or phrases). Insert more paper as needed.

Class Notes/Examples	Your Notes

Answers: **1)** inequality symbols   **2)** rational numbers, real numbers   **3)** negative

Section 1.2 Symbols and Sets of Numbers

Class Notes (continued)	Your Notes

(Insert additional paper as needed.)

Section 1.2 Symbols and Sets of Numbers

Practice:

☐   Complete the Vocabulary, Readiness & Video Check on page 13.

☐   Next, complete any incomplete exercises below. Check and correct your work using the answers and references at the end of this section.

Review this example:	**Your turn:**
1. Insert <, >, or = in the space between each pair of numbers to make each statement true.	2. Insert <, >, or = in the appropriate space to make the statement true.

Review this example:
1. Insert <, >, or = in the space between each pair of numbers to make each statement true.

   a. 2   3

   b. 7   4

   c. 72   27

Your turn:
2. Insert <, >, or = in the appropriate space to make the statement true.

   a. 7   3

   b. 0   7

a. 2 $<$ 3 since 2 is to the left of 3 on the number line.

b. 7 $>$ 4 since 7 is to the right of 4 on the number line.

c. 72 $>$ 27 since 72 is to the right of 27 on the number line.

Review this example:
3. Translate each sentence into a mathematical statement.

   a. Nine   is less than or equal to   eleven.
       ↓   ↓   ↓
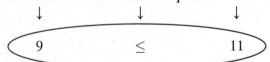

   b. Eight   is greater than   one.
       ↓   ↓   ↓
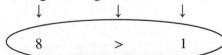

   c. Three   is not equal to   four.
       ↓   ↓   ↓

Your turn:
4. Write each sentence as a mathematical statement.

   a. Five is greater than or equal to four.

   b. Fifteen is not equal to negative two.

Section 1.2 Symbols and Sets of Numbers

Review this example:

5. Given the set $\left\{-2, 0, \frac{1}{4}, -1.5, 112, -3, 11, \sqrt{2}\right\}$, list the numbers in this set that belong to the set of:

a. Natural numbers: (11 and 112)

b. Whole numbers: (0, 11, and 112)

c. Integers: (−3, −2, 0, 11, and 112)

d. Rational numbers: $\left(-3, -2, , -1.5, 0, \frac{1}{4}, 11, \text{ and } 112\right)$

e. Irrational numbers: $\left(\sqrt{2}\right)$

f. Real numbers: (all numbers in the set)

Your turn:

6. Tell which set or sets the number $\frac{2}{3}$ belongs to: natural numbers, whole numbers, integers, rational numbers, irrational numbers, and real numbers.

Review this example:

7. Find the absolute value of each number.

a. $|4| = (4)$ since 4 is 4 units from 0 on a number line.

b. $|-5| = (5)$ since −5 is 5 units from 0 on a number line.

Your turn:

8. Insert <, >, or = in the appropriate space to make the statement true.

a. $|-5|$ -4

b. $|0|$ $|-8|$

	Answer	Text Ref	Video Ref		Answer	Text Ref	Video Ref
1	a. b. > c. >	Ex 1, p. 8		**5**	a. 11, 112 b. 0, 11, 112 c. $-3-2, 0, 11, 112$ d. $-3, -2, -1.5, 0, \frac{1}{4}, 11, 112$ e. $\sqrt{2}$ f. all numbers in the set	Ex 5, p. 11	
2	a. > b. <		Sec 1.2, 1–2/9	**6**	rational, real		Sec 1.2, 7/9
3	a. $9 \le 11$ b. $8 > 1$ c. $3 \ne 4$	Ex 3, p. 8		**7**	a. 4 b. 5	Ex 7a, b, p. 13	
4	a. $5 \ge 4$ b. $15 \ne -2$		Sec 1.2, 4–5/9	**8**	a. > b. <		Sec 1.2, 8–9/9

☐   **Next, insert your homework.** Make sure you attempt all exercises asked of you and show all work, as in the exercises above. Check your answers if possible. Clearly mark any exercises you were unable to correctly complete so that you may ask questions later. DO NOT ERASE YOUR INCORRECT WORK. THIS IS HOW WE UNDERSTAND AND EXPLAIN TO YOU YOUR ERRORS.

Section 1.3 Fractions and Mixed Numbers

Before Class:

☐   Read the objectives on page 16.

☐   Complete the exercises:

1.   When is a fraction said to be in lowest terms?

2.   Write the first five prime numbers.

3.   Read the Fundamental Principle of Fractions box on page 17. If $\frac{a}{b}$ is a fraction and c is a

nonzero real number, then $\dfrac{a \cdot c}{b \cdot c} =$ _____ .

During Class:

☐   **Write your class notes.** Neatly write down **all** examples shown as well as key terms or phrases with definitions. If not applicable or if you were absent, watch the Lecture Series (DVD) for this section and do the same (write down the examples shown as well as key terms or phrases). Insert more paper as needed.

Class Notes/Examples	**Your Notes**

Answers: **1)** when the numerator and the denominator have no factors in common other than 1

2) 2, 3, 5, 7, 11   **3)** $\frac{a}{b}$

Section 1.3 Fractions and Mixed Numbers

Class Notes (continued)

Your Notes

(Insert additional paper as needed.)

Section 1.3 Fractions and Mixed Numbers

Practice:

☐   Complete the Vocabulary, Readiness & Video Check on page 22.

☐   Next, complete any incomplete exercises below. Check and correct your work using the answers and references at the end of this section.

Review this example:

1. Write $\dfrac{42}{49}$ in lowest terms.

Write the numerator and the denominator as products of primes; then apply the fundamental principle to the common factor 7.

$$\frac{42}{49} = \frac{2 \cdot 3 \cdot 7}{7 \cdot 7} = \frac{2 \cdot 3}{7} = \boxed{\frac{6}{7}}$$

Your turn:

2. Write $\dfrac{10}{15}$ in lowest terms.

Review this example:

3. Multiply $\dfrac{2}{15}$ and $\dfrac{5}{13}$. Write the product in lowest terms.

$$\frac{2}{15} \cdot \frac{5}{13} = \frac{2 \cdot 5}{15 \cdot 13} \qquad \begin{array}{l}\text{Multiply numerators.}\\ \text{Multiply denominators.}\end{array}$$

$$= \frac{2 \cdot 5}{3 \cdot 5 \cdot 13} \qquad \begin{array}{l}\text{Write 15 as a product of}\\ \text{primes.}\end{array}$$

$$= \boxed{\frac{2}{39}} \qquad \text{Simplify.}$$

Your turn:

4. Multiply and write the answer in lowest terms.

$$\frac{2}{3} \cdot \frac{3}{4}$$

Review this example:

5. Divide. Write all quotients in lowest terms.

   a. $\dfrac{4}{5} \div \dfrac{5}{16}$ b. $\dfrac{7}{10} \div 14$ c. $\dfrac{3}{8} \div \dfrac{3}{10}$

a. $\dfrac{4}{5} \div \dfrac{5}{16} = \dfrac{4}{5} \cdot \dfrac{16}{5} = \dfrac{4 \cdot 16}{5 \cdot 5} = \boxed{\dfrac{64}{25}}$

b. $\dfrac{7}{10} \div 14 = \dfrac{7}{10} \div \dfrac{14}{1} = \dfrac{7}{10} \cdot \dfrac{1}{14} = \dfrac{7 \cdot 1}{2 \cdot 5 \cdot 2 \cdot 7} = \boxed{\dfrac{1}{20}}$

c. $\dfrac{3}{8} \div \dfrac{3}{10} = \dfrac{3}{8} \cdot \dfrac{10}{3} = \dfrac{3 \cdot 2 \cdot 5}{2 \cdot 2 \cdot 2 \cdot 3} = \boxed{\dfrac{5}{4}}$

Your turn:

6. Divide and write the answer in lowest terms.

$$\frac{3}{4} \div \frac{1}{20}$$

Section 1.3 Fractions and Mixed Numbers

Review this example:

7. Add and write the answer in lowest terms.

$$\frac{3}{10} + \frac{2}{10}$$

$$\frac{3}{10} + \frac{2}{10} = \frac{3+2}{10} = \frac{5}{10} = \frac{5}{2 \cdot 5} = \boxed{\frac{1}{2}}$$

Your turn:

8. Subtract and write the answer in lowest terms.

$$\frac{17}{21} - \frac{10}{21}$$

Review this example:

9. Add and write the answer in lowest terms.

$$\frac{2}{5} + \frac{1}{4}$$

Since 20 is the smallest number that both 5 and 4 divide into evenly, 20 is the least common denominator. Write both fractions as equivalent fractions with denominators of 20. Since

$$\frac{2}{5} \cdot \frac{4}{4} = \frac{2 \cdot 4}{5 \cdot 4} = \frac{8}{20} \quad \text{and} \quad \frac{1}{4} \cdot \frac{5}{5} = \frac{1 \cdot 5}{4 \cdot 5} = \frac{5}{20}$$

then $\dfrac{2}{5} + \dfrac{1}{4} = \dfrac{8}{20} + \dfrac{5}{20} = \boxed{\dfrac{13}{20}}$

Your turn:

10. Subtract and write the answer in lowest terms.

$$\frac{10}{3} - \frac{5}{21}$$

	Answer	Text Ref	Video Ref		Answer	Text Ref	Video Ref
1	$\frac{6}{7}$	Ex 2a, pp. 17–18		6	15		Sec 1.3, 3/7
2	$\frac{2}{3}$		Sec 1.3, 1/7	7	$\frac{1}{2}$	Ex 5b, p. 19	
3	$\frac{2}{39}$	Ex 3, p. 18		8	$\frac{1}{3}$		Sec 1.3, 4/7
4	$\frac{1}{2}$		Sec 1.3, 2/7	9	$\frac{13}{20}$	Ex 7a, p. 20	
5	a. $\frac{64}{25}$ b. $\frac{1}{20}$ c. $\frac{5}{4}$	Ex 4, p. 19		10	$\frac{65}{21}$		Sec 1.3, 6/7

☐ **Next, insert your homework.** Make sure you attempt all exercises asked of you and show all work, as in the exercises above. Check your answers if possible. Clearly mark any exercises you were unable to correctly complete so that you may ask questions later. DO NOT ERASE YOUR INCORRECT WORK. THIS IS HOW WE UNDERSTAND AND EXPLAIN TO YOU YOUR ERRORS.

Section 1.4 Exponents, Order of Operations, Variable Expressions, and Equations

Before Class:

☐   Read the objectives on page 25.

☐   Read the **Helpful Hint** boxes on pages 26, 27, 28, 29, 30, and 31.

☐   Complete the exercises:

1.   In the expression 6^4, what is the base? _____ What is the exponent? _____

2.   Read the Order of Operations box on page 26. If there are no grouping symbols or fraction bars, what is the first thing we should do to simplify an expression?

3.   What is the difference between an expression and an equation?

During Class:

☐   **Write your class notes.** Neatly write down **all** examples shown as well as key terms or phrases with definitions. If not applicable or if you were absent, watch the Lecture Series (DVD) for this section and do the same (write down the examples shown as well as key terms or phrases). Insert more paper as needed.

Class Notes/Examples	**Your Notes**

Answers: **1)** 6, 4   **2)** Evaluate exponential expressions, if any.   **3)** An equation has an equal symbol, an expression does not

Section 1.4 Exponents, Order of Operations, Variable Expressions, and Equations

Class Notes (continued)	**Your Notes**

(Insert additional paper as needed.)

Section 1.4 Exponents, Order of Operations, Variable Expressions, and Equations

Practice:

☐ Complete the Vocabulary, Readiness & Video Check on page 32.

☐ Next, complete any incomplete exercises below. Check and correct your work using the answers and references at the end of this section.

Review this example:	**Your turn:**
1. Simplify the expression.	**2.** Simplify the expression.
$6 \div 3 + 5^2$	$5 + 6 \cdot 2$
$6 \div 3 + 5^2 = 6 \div 3 + 25$ Evaluate 5^2 first.	
$= 2 + 25$ Divide.	
$= \boxed{27}$ Add.	

Review this example:	**Your turn:**
3. Simplify: $\dfrac{3 + |4 - 3| + 2^2}{6 - 3}$	**4.** Simplify the expression.
	$\dfrac{|6 - 2| + 3}{8 + 2 \cdot 5}$
The fraction bar and the absolute value bars serve as grouping symbols. Simplify each separately.	
$\dfrac{3 + |4 - 3| + 2^2}{6 - 3} = \dfrac{3 + |1| + 2^2}{6 - 3}$	
$= \dfrac{3 + 1 + 2^2}{3}$	
$= \dfrac{3 + 1 + 4}{3}$	
$= \boxed{\dfrac{8}{3}}$	

Review this example:	**Your turn:**
5. Evaluate the expression if $x = 3$ and $y = 2$.	**6.** Evaluate the expression if $x = 12$, $y = 8$, and $z = 4$.
$\dfrac{x}{y} + \dfrac{y}{2}$	$\dfrac{x^2 + z}{y^2 + 2z}$
Replace x with 3 and y with 2.	
$\dfrac{x}{y} + \dfrac{y}{2} = \dfrac{3}{2} + \dfrac{2}{2} = \boxed{\dfrac{5}{2}}$	

Section 1.4 Exponents, Order of Operations, Variable Expressions, and Equations

Review this example:

7. Decide whether 2 is a solution of
   $3x + 10 = 8x$.

   Replace x with 2 and see if a true statement results.
   $3x + 10 = 8x$

   $3(2) + 10 \overset{?}{=} 8(2)$

   $6 + 10 \overset{?}{=} 16$

   $16 = 16$

   Since we arrived at a true statement after replacing x with 2 and simplifying both sides of the equation,

   2 is a solution of the equation.

Your turn:

8. Is 0 a solution of $x = 5x + 15$?

Review this example:

9. Write the sentence as an equation or inequality. Let x represent the unknown number.

   Four times a number, added to 17, is not equal to 21.

Four times a number	added to	17	is not equal to	21
↓	↓	↓	↓	↓
$4x$	$+$	17	$\neq$	21

Your turn:

10. Write the sentence as an equation or inequality.

    One increased by two equals the quotient of nine and three.

	Answer	Text Ref	Video Ref		Answer	Text Ref	Video Ref
1	27	Ex 2a, p. 26		**6**	$\dfrac{37}{18}$		Sec 1.4, 7/12
2	17		Sec 1.4, 3/12	**7**	yes	Ex 7, p. 30	
3	$\dfrac{8}{3}$	Ex 3, p. 27		**8**	no		Sec 1.4, 9/12
4	$\dfrac{7}{18}$		Sec 1.4, 5/12	**9**	$4x + 17 \neq 21$	Ex 9c, p. 31	
5	$\dfrac{5}{2}$	Ex 6c, p. 29		**10**	$1 + 2 = 9 \div 3$		Sec 1.4, 11/12

☐   **Next, insert your homework.** Make sure you attempt all exercises asked of you and show all work, as in the exercises above. Check your answers if possible. Clearly mark any exercises you were unable to correctly complete so that you may ask questions later. DO NOT ERASE YOUR INCORRECT WORK. THIS IS HOW WE UNDERSTAND AND EXPLAIN TO YOU YOUR ERRORS.

Section 1.5 Adding Real Numbers

Before Class:

☐ Read the objectives on page 35.

☐ Read the **Helpful Hint** boxes on page 38.

☐ Complete the exercises:

1. How do we add two negative numbers? What is the sign of the sum?

2. How do we add two numbers with different signs? What is the sign of the sum?

During Class:

☐ **Write your class notes.** Neatly write down **all** examples shown as well as key terms or phrases with definitions. If not applicable or if you were absent, watch the Lecture Series (DVD) for this section and do the same (write down the examples shown as well as key terms or phrases). Insert more paper as needed.

Class Notes/Examples	Your Notes

Answers: **1)** Add their absolute values, negative   **2)** Subtract the smaller absolute value from the larger absolute value, use the sign of the number whose absolute value is larger.

Section 1.5 Adding Real Numbers

Class Notes (continued)	Your Notes

(Insert additional paper as needed.)

Section 1.5 Adding Real Numbers

Practice:

☐   Complete the Vocabulary, Readiness & Video Check on page 40.

☐   Next, complete any incomplete exercises below. Check and correct your work using the
      answers and references at the end of this section.

Review this example:	**Your turn:**
1. Add: $-3+(-7)$	**2.** Add: $-2+(-3)$

Notice that the numbers have the same sign. Add
their absolute values and use their common sign.

$-3+(-7)=\boxed{-10}$

Review this example:	**Your turn:**
3. Add: $3+(-7)$	**4.** Add: $5+(-7)$

Notice that the numbers have different signs.
Subtract their absolute values. The negative
number, -7, has the larger absolute value so the
sum is negative.

$3+(-7)=\boxed{-4}$

Review this example:	**Your turn:**
5. Add: $3+(-7)+(-8)$	**6.** Add: $-21+(-16)+(-22)$

Perform the additions from left to right.

$3+(-7)+(-8)=-4+(-8)$
$\qquad\qquad\quad\ =\boxed{-12}$

Section 1.5 Adding Real Numbers

Review this example:

7. In Philadelphia, Pennsylvania, the record extreme high temperature is 104°F. Decrease this temperature by 111 degrees, and the result is the record extreme low temperature. Find this temperature.

Extreme low temperature	=	Extreme high temperature	+	Decrease of 111°
↓		↓		↓

$$\text{Extreme low temperature} = 104 + -111$$
$$= -7$$

The record extreme low temperature in Philadelphia, Pennsylvania is $-7°F$.

Your turn:

8. The low temperature in Anoka, Minnesota, was −15° last night. During the day it rose only 9°. Find the high temperature for the day.

Review this example:

9. Find the opposite or additive inverse of each number.

   a. 5   b. −6

a. The opposite of 5 is -5.
   Notice that 5 and −5 are on opposite sides of 0 when plotted on a number line and are equal distances away.

b. The opposite of −6 is 6.

Your turn:

10. Find each additive inverse or opposite.

   a. 6   b. −2

	Answer	Text Ref	Video Ref			Answer	Text Ref	Video Ref
1	−10	Ex 3a, p. 36			**6**	−59		Sec 1.5, 6/13
2	−5		Sec 1.5, 2/13		**7**	−7°F	Ex 8, p. 39	
3	−4	Ex 5a, p. 37			**8**	−6°		Sec 1.5, 9/13
4	−2		Sec 1.5, 4/13		**9**	a. −5 b. 6	Ex 9a, b, p. 39	
5	−12	Ex 7a, p. 38			**10**	a. −6 b. 2		Sec 1.5, 10−11/13

☐   **Next, insert your homework.** Make sure you attempt all exercises asked of you and show all work, as in the exercises above. Check your answers if possible. Clearly mark any exercises you were unable to correctly complete so that you may ask questions later. DO NOT ERASE YOUR INCORRECT WORK. THIS IS HOW WE UNDERSTAND AND EXPLAIN TO YOU YOUR ERRORS.

Section 1.6 Subtracting Real Numbers

Before Class:

☐   Read the objectives on page 43.

☐   Read the **Helpful Hint** boxes on pages 43 and 46.

☐   Complete the exercises:

1.   To find the difference of two numbers, add the first number to the

_____ of the second number.

2.   Two angles are complementary if their sum is _____ .

3.   Two angles are supplementary if their sum is _____ .

During Class:

☐   **Write your class notes.** Neatly write down **all** examples shown as well as key terms or phrases with definitions. If not applicable or if you were absent, watch the Lecture Series (DVD) for this section and do the same (write down the examples shown as well as key terms or phrases). Insert more paper as needed.

Class Notes/Examples	Your Notes

Answers: **1)** opposite   **2)** 90°   **3)** 180°

Section 1.6 Subtracting Real Numbers

Class Notes (continued)	Your Notes

(Insert additional paper as needed.)

Section 1.6 Subtracting Real Numbers

Practice:

☐   Complete the Vocabulary, Readiness & Video Check on page 47.

☐   Next, complete any incomplete exercises below. Check and correct your work using the answers and references at the end of this section.

Review this example:	**Your turn:**
1. Subtract: $-\dfrac{3}{10}-\dfrac{5}{10}$	**2.** Subtract: $-\dfrac{3}{11}-\left(-\dfrac{5}{11}\right)$

$$-\frac{3}{10}-\frac{5}{10}=-\frac{3}{10}+\left(-\frac{5}{10}\right)=-\frac{8}{10}=\boxed{-\frac{4}{5}}$$

Review this example:	**Your turn:**
3. Simplify: $2^3-\left|10\right|+\left[-6-(-5)\right]$	**4.** Simplify: $\left|-3\right|+2^2+\left[-4-(-6)\right]$
	(Remember the order of operations.)

Start simplifying the expression inside the brackets by writing $-6-(-5)$ as a sum.

$$2^3-\left|10\right|+\left[-6-(-5)\right]=2^3-\left|10\right|+\left[-6+5\right]$$
$$=2^3-\left|10\right|+\left[-1\right]$$
$$=8-10+(-1)$$
$$=8+(-10)+(-1)$$
$$=-2+(-1)$$
$$=\boxed{-3}$$

Review this example:	**Your turn:**
5. Find the value of the expression $\dfrac{x-y}{12+x}$ when $x=2$ and $y=-5$.	**6.** Evaluate the expression $\dfrac{9-x}{y+6}$ when $x=-5$ and $y=4$.

Replace x with 2 and y with -5. Be sure to put parentheses around -5 to separate signs. Then simplify the resulting expression.

$$\frac{x-y}{12+x}=\frac{2-(-5)}{12+2}=\frac{2+5}{14}=\frac{7}{14}=\boxed{\frac{1}{2}}$$

Section 1.6 Subtracting Real Numbers

Review this example:

7. The highest point in the United States is the top of Mount McKinley, at a height of 20,320 feet above sea level. The lowest point is Death Valley, California, which is 282 feet below sea level. How much higher is Mount McKinley than Death Valley?

To find the variation in elevation between the two heights, find the difference of the high point and the low point.

$$20,320 - (-282) = 20,320 + 282$$
$$= 20,602$$

The variation in elevation is 20,602 feet.

Your turn:

8. A commercial jet liner hits an air pocket and drops 250 feet. After climbing 120 feet, it drops another 178 feet. What is its overall vertical change?

Review this example:

9. Find the unknown angle.

These angles are complementary, so their sum is 90°. This means that x is $90° - 38°$.

$$x = 90° - 38° = 52°$$

Your turn:

10. Find the unknown angle.

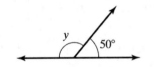

	Answer	Text Ref	Video Ref		Answer	Text Ref	Video Ref
1	$-\dfrac{4}{5}$	Ex 2b, p. 44		6	$\dfrac{7}{5}$		Sec 1.6, 7/10
2	$\dfrac{2}{11}$		Sec 1.6, 3/10	7	20,602 ft	Ex 7, p.46	
3	-3	Ex 5b, p. 45		8	-308 ft		Sec 1.6, 8/10
4	9		Sec 1.6, 6/10	9	52°	Ex 8a, p. 47	
5	$\dfrac{1}{2}$	Ex 6a, p. 45		10	130°		Sec 1.6, 9/10

☐ **Next, insert your homework.** Make sure you attempt all exercises asked of you and show all work, as in the exercises above. Check your answers if possible. Clearly mark any exercises you were unable to correctly complete so that you may ask questions later. DO NOT ERASE YOUR INCORRECT WORK. THIS IS HOW WE UNDERSTAND AND EXPLAIN TO YOU YOUR ERRORS.

Section 1.7 Multiplying and Dividing Real Numbers

Before Class:

☐ Read the objectives on page 51.

☐ Read the **Helpful Hint** boxes on pages 52 and 53.

☐ Complete the exercises:

1. If we multiply an even number of negative numbers, the product is

    _____ .

2. The base of the exponential expression -4^3 is _____ .

3. Two numbers whose product is 1 are called _____ or

    _____ of each other.

During Class:

☐ **Write your class notes.** Neatly write down **all** examples shown as well as key terms or phrases with definitions. If not applicable or if you were absent, watch the Lecture Series (DVD) for this section and do the same (write down the examples shown as well as key terms or phrases). Insert more paper as needed.

Class Notes/Examples	**Your Notes**

Answers: **1)** positive   **2)** 4   **3)** reciprocals, multiplicative inverses

Section 1.7 Multiplying and Dividing Real Numbers

Class Notes (continued)	**Your Notes**

(Insert additional paper as needed.)

Section 1.7 Multiplying and Dividing Real Numbers

Practice:

☐   Complete the Vocabulary, Readiness & Video Check on page 58.

☐   Next, complete any incomplete exercises below. Check and correct your work using the answers and references at the end of this section.

Review this example:
1. Multiply.

   a. $(-1.2)(0.05)$ b. $\dfrac{2}{3}\cdot\left(-\dfrac{7}{10}\right)$

   c. $\left(-\dfrac{4}{5}\right)(-20)$

  a. $(-1.2)(0.05)=-\left[(1.2)(0.05)\right]=\boxed{-0.06}$

  b. $\dfrac{2}{3}\cdot\left(-\dfrac{7}{10}\right)=-\dfrac{2\cdot7}{3\cdot10}=-\dfrac{2\cdot7}{3\cdot2\cdot5}=\boxed{-\dfrac{7}{15}}$

  c. $\left(-\dfrac{4}{5}\right)(-20)=\dfrac{4\cdot20}{5\cdot1}=\dfrac{4\cdot4\cdot5}{5\cdot1}=\dfrac{16}{1}=\boxed{16}$

Your turn:
2. Multiply.

  a. $2(-1)$

  b. $-5(-10)$

  c. $\dfrac{2}{3}\left(-\dfrac{4}{9}\right)$

Review this example:
3. Evaluate.

   a. $(-2)^3$ b. -2^3

  a. $(-2)^3=(-2)(-2)(-2)=\boxed{-8}$

  b. $-2^3=-(2\cdot2\cdot2)=\boxed{-8}$

Your turn:
4. Evaluate.

  a. $(-2)^4$

  b. -2^4

Review this example:
5. Divide.

   a. $\dfrac{-36}{3}$ b. $\dfrac{2}{3}\div\left(-\dfrac{5}{4}\right)$

  a. $\dfrac{-36}{3}=\boxed{-12}$

  b. $\dfrac{2}{3}\div\left(-\dfrac{5}{4}\right)=\dfrac{2}{3}\cdot\left(-\dfrac{4}{5}\right)=\boxed{-\dfrac{8}{15}}$

Your turn:
6. Divide.

  a. $\dfrac{18}{-2}$

  b. $-\dfrac{5}{9}\div\left(-\dfrac{3}{4}\right)$

Section 1.7 Multiplying and Dividing Real Numbers

Review this example:

7. Simplify: $\dfrac{(-12)(-3)+3}{-7-(-2)}$

$\dfrac{(-12)(-3)+3}{-7-(-2)} = \dfrac{36+3}{-7+2} = \dfrac{39}{-5} = \boxed{-\dfrac{39}{5}}$

Your turn:

8. Simplify: $\dfrac{6-2(-3)}{4-3(-2)}$

Review this example:

9. If $x=-2$ and $y=-4$, evaluate x^4-y^2.

Replace x with -2 and y with -4.

$x^4 - y^2 = (-2)^4 - (-4)^2$

$= 16 - (16)$

$= \boxed{0}$

Your turn:

10. If $x=-5$ and $y=-3$, evaluate

$\dfrac{x^2+y}{3y}$.

	Answer	Text Ref	Video Ref		Answer	Text Ref	Video Ref
1	a. -0.06 b. $-\dfrac{7}{15}$ c. 16	Ex 3, p. 53		6	a. -9 b. $\dfrac{20}{27}$		Sec 1.7, 11/18, 15/18
2	a. -2 b. 50 c. $-\dfrac{8}{27}$		Sec 1.7, 3–4/18, 6/18	7	$-\dfrac{39}{5}$	Ex 9a, p. 56	
3	a. -8 b. -8	Ex 4a, b, p. 53		8	$\dfrac{6}{5}$		Sec 1.7, 16/18
4	a. 16 b. -16		Sec 1.7, 7–8/18	9	0	Ex 10b, p. 57	
5	a. -12 b. $-\dfrac{8}{15}$	Ex 7b, c, p. 55		10	$-\dfrac{22}{9}$		Sec 1.7, 17/18

☐   **Next, insert your homework.** Make sure you attempt all exercises asked of you and show all work, as in the exercises above. Check your answers if possible. Clearly mark any exercises you were unable to correctly complete so that you may ask questions later. DO NOT ERASE YOUR INCORRECT WORK. THIS IS HOW WE UNDERSTAND AND EXPLAIN TO YOU YOUR ERRORS.

Section 1.8 Properties of Real Numbers

Before Class:

☐   Read the objectives on page 61.

☐   Read the **Helpful Hint** boxes on pages 61, 62, and 63.

☐   Complete the exercises:

1.   What is the identity element for addition?

2.   What is the identity element for multiplication?

3.   The product of a real number and its multiplicative inverse is _____ .

4.   The sum of a real number and its additive inverse is _____ .

During Class:

☐   **Write your class notes.** Neatly write down **all** examples shown as well as key terms or phrases with definitions. If not applicable or if you were absent, watch the Lecture Series (DVD) for this section and do the same (write down the examples shown as well as key terms or phrases). Insert more paper as needed.

Class Notes/Examples	**Your Notes**

Answers: **1)** 0   **2)** 1   **3)** 1   **4)** 0

Section 1.8 Properties of Real Numbers

| Class Notes (continued) | Your Notes |

(Insert additional paper as needed.)

Section 1.8 Properties of Real Numbers

Practice:

☐   Complete the Vocabulary, Readiness & Video Check on page 66.

☐   Next, complete any incomplete exercises below. Check and correct your work using the
       answers and references at the end of this section.

Review this example:	**Your turn:**
1. Use a commutative property to complete each statement.	2. Use a commutative property to complete each statement.
a. $x+5=\boxed{5+x}$ by the commutative property of addition	a. $x+16=$ _____
b. $3\cdot x=\boxed{x\cdot 3}$ by the commutative property of multiplication	b. $xy=$ _____

Review this example:	**Your turn:**
3. Use an associative property to complete each statement.	4. Use an associative property to complete each statement.
a. $5+(4+6)=\boxed{(5+4)+6}$ by the associative property of addition	a. $(xy)\cdot z=$ _____
b. $(-1\cdot 2)\cdot 5=\boxed{-1\cdot(2\cdot 5)}$ by the associative property of multiplication	b. $(a+b)+c=$ _____

Review this example:	**Your turn:**
5. Simplify each expression.	6. Use the commutative and associative properties to simplify each expression.
a. $10+(x+12)=10+(12+x)=(10+12)+x$ $=\boxed{22+x}$	a. $8+(9+b)$
b. $-3(7x)=(-3\cdot 7)x=\boxed{-21x}$	b. $4(6y)$

Review this example:	**Your turn:**
7. Use the distributive property to write each expression without parentheses. Then simplify if possible.	8. Use the distributive property to write each expression without parentheses. Then simplify the result.
a. $-5(-3+2z)=-5(-3)+(-5)(2z)=\boxed{15-10z}$	a. $3(6+x)$
b. $5(x+3y-z)=5(x)+5(3y)-5(z)$ $=\boxed{5x+15y-5z}$	b. $-(r-3-7p)$

27

Section 1.8 Properties of Real Numbers

Review this example:

9. Name the property illustrated by each true statement.

a. $3 \cdot y = y \cdot 3$

    commutative property of multiplication

b. $(x + 7) + 9 = x + (7 + 9)$

    associative property of addition

c. $(b + 0) + 3 = b + 3$

    identity element of addition

d. $-2 \cdot \left(-\dfrac{1}{2}\right) = 1$

    multiplicative inverse property

Your turn:

10. Name the property illustrated by each true statement.

a. $1 \cdot 9 = 9$

b. $6 \cdot \dfrac{1}{6} = 1$

c. $0 + 6 = 6$

d. $(11 + r) + 8 = (r + 11) + 8$

	Answer	Text Ref	Video Ref		Answer	Text Ref	Video Ref
1	a. $5 + x$ b. $x \cdot 3$	Ex 1, p. 61		**6**	a. $17 + b$ b. $24y$		Sec 1.8, 6–7/15
2	a. $16 + x$ b. yx		Sec 1.8, 1–2/15	**7**	a. $15 - 10z$ b. $5x + 15y - 5z$	Ex 4b, c, p. 63	
3	a. $(5 + 4) + 6$ b. $-1 \cdot (2 \cdot 5)$	Ex 2, p. 62		**8**	a. $18 + 3x$ b. $-r + 3 + 7p$		Sec 1.8, 8–9/15
4	a. $x \cdot (yz)$ b. $a + (b + c)$		Sec 1.8, 4–5/15	**9**	a. commutative property of multiplication b. associative property of addition c. identity element of addition d. multiplicative inverse property	Ex 6a, b, c, e, p. 65	
5	a. $22 + x$ b. $-21x$	Ex 3, p. 62		**10**	a. identity element of multiplication b. multiplicative inverse property c. identity element of addition d. commutative property of addition		Sec 1.8, 12–15/15

☐   **Next, insert your homework.** Make sure you attempt all exercises asked of you and show all work, as in the exercises above. Check your answers if possible. Clearly mark any exercises you were unable to correctly complete so that you may ask questions later. DO NOT ERASE YOUR INCORRECT WORK. THIS IS HOW WE UNDERSTAND AND EXPLAIN TO YOU YOUR ERRORS.

Preparing for the Chapter 1 Test

Start preparing for your Chapter 1 Test as soon as possible. Pay careful attention to any instructor discussion about this test, especially discussion on what sections you will be responsible for, etc.

☐ Work the Chapter 1 Vocabulary Check on page 68.

☐ Read your Class Notes/Examples for each section covered on your Chapter 1 Test. Look for any unresolved questions you may have.

☐ Complete as many of the Chapter 1 Review exercises as possible (page 72). Remember, the odd answers are in the back of your text.

☐ **Most important:** Place yourself in "test" conditions (see below) and work the Chapter 1 Test (page 75) as a practice test the day before your actual test. To honestly assess how you are doing, try the following:
- Work on a few blank sheets of paper.
- Give yourself the same amount of time you will be given for your actual test.
- Complete this Chapter 1 Practice Test without using your notes or your text.
- If you have any time left after completing this practice test, check your work and try to find any errors on your own.
- Once done, use the back of your book to check ALL answers.
- Try to correct any errors on your own.
- Use the Chapter Test Prep Video (CTPV) to correct any errors you were unable to correct on your own. You can find these videos in the Interactive DVD Lecture Series, in MyMathLab, and on YouTube. Search Martin-Gay Beginning Algebra and click "Channels."

I wish you the best of luck….Elayn Martin-Gay

Section 2.1 Simplifying Algebraic Expressions

Before Class:

☐ Read the objectives on page 77.

☐ Read the **Helpful Hint** boxes on pages 77, 78, and 80.

☐ Complete the exercises:

1. To combine like terms, add the _____ and multiply the result by the common variable factors.

2. An algebraic expression containing the sum or difference of like terms can be simplified

    by applying the _____ property.

3. Like terms have the same _____ raised to exactly the same

    _____.

During Class:

☐ **Write your class notes.** Neatly write down **all** examples shown as well as key terms or phrases with definitions. If not applicable or if you were absent, watch the Lecture Series (DVD) for this section and do the same (write down the examples shown as well as key terms or phrases). Insert more paper as needed.

Class Notes/Examples	Your Notes

Answers: **1)** numerical coefficients   **2)** distributive   **3)** variables, powers

Section 2.1 Simplifying Algebraic Expressions

Class Notes (continued)	**Your Notes**

(Insert additional paper as needed.)

Copyright © 2013 Pearson Education, Inc.

Section 2.1 Simplifying Algebraic Expressions

Practice:

☐   Complete the Vocabulary, Readiness & Video Check on page 82.

☐   Next, complete any incomplete exercises below. Check and correct your work using the answers and references at the end of this section.

Review this example:	**Your turn:**
1. Simplify each expression by combining like terms.	**2.** Simplify the expression by combining any like terms.
a. $7x - 3x$ b. $10y^2 + y^2$ c. $8x^2 + 2x - 3x$	$3x + 2x$
a. $7x - 3x = (7-3)x = \boxed{4x}$	
b. $10y^2 + y^2 = 10y^2 + 1y^2 = (10+1)y^2 = \boxed{11y^2}$	
c. $8x^2 + 2x - 3x = 8x^2 + (2-3)x = \boxed{8x^2 - x}$	

Review this example:	**Your turn:**
3. Simplify by combining like terms.	**4.** Simplify the expression by combining any like terms.
$-5a - 3 + a + 2$	$8x^3 + x^3 - 11x^3$
$-5a - 3 + a + 2 = -5a + 1a + (-3+2)$	
$\qquad = (-5+1)a + (-3+2)$	
$\qquad = \boxed{-4a - 1}$	

Review this example:	**Your turn:**
5. Simplify the following expression.	**6.** Simplify the expression. First use the distributive property to remove any parentheses.
$-2(4x+7) - (3x-1)$	$5(x+2) - (3x-4)$
$-2(4x+7) - (3x-1) = -8x - 14 - 3x + 1$	
$\qquad\qquad = \boxed{-11x - 13}$	

Section 2.1 Simplifying Algebraic Expressions

Review this example:

7. Write the following phrase as an algebraic expression. Then simplify if possible.

"Subtract $4x - 2$ from $2x - 3$."

"Subtract $4x - 2$ from $2x - 3$" translates to $(2x - 3) - (4x - 2)$. Next, simplify the algebraic expression.

$$(2x - 3) - (4x - 2) = 2x - 3 - 4x + 2$$
$$= \boxed{-2x - 1}$$

Your turn:

8. Write the following as an algebraic expression. Simplify if possible.

"Subtract $5m - 6$ from $m - 9$."

Review this example:

9. Write the following phrase as an algebraic expression and simplify if possible. Let x represent the unknown number.

"The sum of twice a number, 3 times the number, and 5 times the number"

The phrase "the sum of" means we add.

twice a number	added to	3 times the number	added to	5 times the number
↓	↓	↓	↓	↓
$2x$	$+$	$3x$	$+$	$5x$

Now simplify: $2x + 3x + 5x = \boxed{10x}$

Your turn:

10. Write the following phrase as an algebraic expression and simplify if possible. Let x represent the unknown number.

"The sum of 5 times a number and -2, added to 7 times a number"

	Answer	Text Ref	Video Ref		Answer	Text Ref	Video Ref
1	a. $4x$ b. $11y^2$ c. $8x^2 - x$	Ex 3, p. 78		**6**	$2x + 14$		Sec 2.1, 11/14
2	$5x$		Sec 2.1, 8/14	**7**	$-2x - 1$	Ex 7, p. 80	
3	$-4a - 1$	Ex 4b, p. 79		**8**	$-4m - 3$		Sec 2.1, 12/14
4	$-2x^3$		Sec 2.1, 9/14	**9**	$10x$	Ex 8d, p. 81	
5	$-11x - 13$	Ex 6b, p. 80		**10**	$-2 + 12x$		Sec 2.1, 14/14

☐ **Next, insert your homework.** Make sure you attempt all exercises asked of you and show all work, as in the exercises above. Check your answers if possible. Clearly mark any exercises you were unable to correctly complete so that you may ask questions later. DO NOT ERASE YOUR INCORRECT WORK. THIS IS HOW WE UNDERSTAND AND EXPLAIN TO YOU YOUR ERRORS.

Section 2.2 The Addition Property of Equality

Before Class:

☐ Read the objectives on page 85.

☐ Read the **Helpful Hint** box on page 87.

☐ Complete the exercises:

   1. The _____ property of _____
      guarantees that adding the same number to both sides of an equation does not change the
      solution of the equation.

   2. To solve an equation, does it matter which side of the equal sign the variable is on?

During Class:

☐ **Write your class notes.** Neatly write down **all** examples shown as well as key terms or
   phrases with definitions. If not applicable or if you were absent, watch the Lecture Series
   (DVD) for this section and do the same (write down the examples shown as well as key terms
   or phrases). Insert more paper as needed.

Class Notes/Examples	**Your Notes**

Answers: **1)** addition, equality   **2)** no

Section 2.2 The Addition Property of Equality

Class Notes (continued)

Your Notes

(Insert additional paper as needed.)

Section 2.2 The Addition Property of Equality

Practice:

☐ Complete the Vocabulary, Readiness & Video Check on page 90.

☐ Next, complete any incomplete exercises below. Check and correct your work using the answers and references at the end of this section.

Review this example:	**Your turn:**
1. Solve $5t - 5 = 6t + 2$ for t.	**2.** Solve: $3x - 6 = 2x + 5$

We first want all terms containing t on one side of the equation and all other terms on the other side of the equation. To do this, first subtract $5t$ from both sides of the equation.

$$5t - 5 = 6t + 2$$
$$5t - 5 - 5t = 6t + 2 - 5t$$
$$-5 = t + 2$$

Next, subtract 2 from both sides.

$$-5 = t + 2$$
$$-5 - 2 = t + 2 - 2$$
$$-7 = t$$

Check the solution, -7, in the original equation. The solution is $\boxed{-7.}$

Review this example:	**Your turn:**
3. Solve: $2x + 3x - 5 + 7 = 10x + 3 - 6x - 4$	**4.** Solve: $13x - 9 + 2x - 5 = 12x - 1 + 2x$

First simplify both sides of the equation.

$$2x + 3x - 5 + 7 = 10x + 3 - 6x - 4$$
$$5x + 2 = 4x - 1$$

Next, we want all terms with a variable on one side of the equation and all numbers on the other side.

$$5x + 2 - 4x = 4x - 1 - 4x$$
$$x + 2 = -1$$
$$x + 2 - 2 = -1 - 2$$
$$x = -3$$

Check:
$$2x + 3x - 5 + 7 = 10x + 3 - 6x - 4$$
$$2(-3) + 3(-3) - 5 + 7 \overset{?}{=} 10(-3) + 3 - 6(-3) - 4$$
$$-6 - 9 - 5 + 7 \overset{?}{=} -30 + 3 + 18 - 4$$
$$-13 = -13$$

The solution is $\boxed{-3.}$

Section 2.2 The Addition Property of Equality

Review this example:

5. Solve: $6(2a-1)-(11a+6)=7$

$$6(2a-1)-1(11a+6)=7$$
$$6(2a)+6(-1)-1(11a)-1(6)=7$$
$$12a-6-11a-6=7$$
$$a-12=7$$
$$a-12+12=7+12$$
$$a=\boxed{19}$$

Check by replacing a with 19 in the original equation.

Your turn:

6. Solve: $15-(6-7k)=2+6k$

Review this example:

7. The sum of two numbers is 8. If one number is x write an expression representing the other number.

If the sum of two numbers is 8 and one number is x we find the other number by subtracting x from 8. The other number is represented by $\boxed{8-x}$.

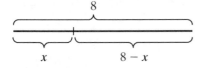

Your turn:

8. Two numbers have a sum of 20. If one number is p, express the other number in terms of p.

	Answer	Text Ref	Video Ref		Answer	Text Ref	Video Ref
1	-7	Ex 4, p. 87		**5**	19	Ex 6, p. 88	
2	11		Sec 2.2, 4/8	**6**	-7		Sec 2.2, 6/8
3	-3	Ex 5, p. 88		**7**	$8-x$	Ex 8b, p. 89	
4	13		Sec 2.2, 5/8	**8**	$20-p$		Sec 2.2, 7/8

☐   **Next, insert your homework.** Make sure you attempt all exercises asked of you and show all work, as in the exercises above. Check your answers if possible. Clearly mark any exercises you were unable to correctly complete so that you may ask questions later. DO NOT ERASE YOUR INCORRECT WORK. THIS IS HOW WE UNDERSTAND AND EXPLAIN TO YOU YOUR ERRORS.

Section 2.3 The Multiplication Property of Equality

Before Class:

☐   Read the objectives on page 93.

☐   Read the **Helpful Hint** boxes on pages 95 and 98.

☐   Complete the exercises:

1.   What does the multiplication property of equality guarantee?

2.   If x is an odd integer, how can we express the next two consecutive odd integers in terms of x?

During Class:

☐   **Write your class notes.** Neatly write down **all** examples shown as well as key terms or phrases with definitions. If not applicable or if you were absent, watch the Lecture Series (DVD) for this section and do the same (write down the examples shown as well as key terms or phrases). Insert more paper as needed.

Class Notes/Examples	**Your Notes**

Answers: **1)** Multiplying both sides of an equation by the same nonzero number does not change the solution of the equation.   **2)** $x+2$, $x+4$

Section 2.3 The Multiplication Property of Equality

Class Notes (continued)	**Your Notes**

(Insert additional paper as needed.)

Section 2.3 The Multiplication Property of Equality

Practice:

☐   Complete the Vocabulary, Readiness & Video Check on page 98.

☐   Next, complete any incomplete exercises below. Check and correct your work using the answers and references at the end of this section.

Review this example:

1. Solve: $-3x = 33$

To get x alone, we divide both sides by the coefficient of x.

$$-3x = 33 \qquad\qquad \text{Check:} \qquad -3x = 33$$

$$\frac{-3x}{-3} = \frac{33}{-3} \qquad\qquad\qquad -3(-11) \overset{?}{=} 33$$

$$1x = -11 \qquad\qquad\qquad\qquad 33 = 33$$

$$x = -11$$

The solution is $\boxed{-11.}$

Your turn:

2. Solve the equation $-5x = -20$.
Check the solution.

Review this example:

3. Solve: $\dfrac{y}{7} = 20$

Multiply both sides of the equation by 7, the reciprocal of $\dfrac{1}{7}$.

$$\frac{y}{7} = 20 \qquad\qquad \text{Check:} \qquad \frac{y}{7} = 20$$

$$\frac{1}{7}y = 20 \qquad\qquad\qquad \frac{140}{7} \overset{?}{=} 20$$

$$7 \cdot \frac{1}{7}y = 7 \cdot 20 \qquad\qquad\qquad 20 = 20$$

$$1y = 140$$

$$y = 140$$

The solution is $\boxed{140.}$

Your turn:

4. Solve the equation $\dfrac{2}{3}x = -8$. Check the solution.

Section 2.3 The Multiplication Property of Equality

Review this example:

5. Solve: $7x - 3 = 5x + 9$

To get x alone, first use the addition property to get variable terms on one side of the equation and numbers on the other side.

$$7x - 3 = 5x + 9$$

$$7x - 3 - 5x = 5x + 9 - 5x$$

$$2x - 3 = 9$$

$$2x - 3 + 3 = 9 + 3$$

$$2x = 12$$

Use the multiplication property to get x alone.

$$\frac{2x}{2} = \frac{12}{2}$$

$$x = 6$$

To check, replace x with 6 in the original equation to see that a true statement results.

The solution is $\boxed{6.}$

Your turn:

6. Solve the equation $8x + 20 = 6x + 18$.

Review this example:

7. Solve: $5(2x + 3) = -1 + 7$

$$5(2x + 3) = -1 + 7$$

$$5(2x) + 5(3) = -1 + 7$$

$$10x + 15 = 6$$

$$10x + 15 - 15 = 6 - 15$$

$$10x = -9$$

$$\frac{10x}{10} = -\frac{9}{10}$$

$$x = -\frac{9}{10}$$

To check, replace x with $-\frac{9}{10}$ in the original equation to see that a true statement results. The solution is $\boxed{-\frac{9}{10}.}$

Your turn:

8. Solve the equation

$$9(3x + 1) = 4x - 5x.$$

Section 2.3 The Multiplication Property of Equality

Review this example:

9. If x is the first of three consecutive integers, express the sum of the three integers in terms of x. Simplify if possible.

The second consecutive integer is always 1 more than the first, and the third consecutive integer is 2 more than the first. If x is the first of three consecutive integers, the three consecutive integers are

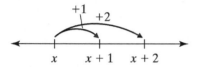

Their sum is $x+(x+1)+(x+2)$, which simplifies to $3x+3$.

Your turn:

10. Write the algebraic expression described. Simplify if possible.

If x is the first of four consecutive integers, express the sum of the first integer and the third integer as an algebraic expression containing the variable x.

	Answer	Text Ref	Video Ref		Answer	Text Ref	Video Ref
1	−11	Ex 2, p. 94		6	−1		Sec 2.3, 5/8
2	4		Sec 2.3, 1/8	7	$-\dfrac{9}{10}$	Ex 9, p. 97	
3	140	Ex 3, p. 94		8	$-\dfrac{9}{28}$		Sec 2.3, 7/8
4	−12		Sec 2.3, 2/8	9	$3x+3$	Ex 10, p. 97	
5	6	Ex 8, p. 96		10	$2x+2$		Sec 2.3, 8/8

☐ **Next, insert your homework.** Make sure you attempt all exercises asked of you and show all work, as in the exercises above. Check your answers if possible. Clearly mark any exercises you were unable to correctly complete so that you may ask questions later. DO NOT ERASE YOUR INCORRECT WORK. THIS IS HOW WE UNDERSTAND AND EXPLAIN TO YOU YOUR ERRORS.

Section 2.3 The Multiplication Property of Equality

Section 2.4 Solving Linear Equations

Before Class:

☐   Read the objectives on page 100.

☐   Read the **Helpful Hint** boxes on pages 101, 102 and 103

☐   Complete the exercises:

  1. Read the Solving Linear Equations in One Variable box on page 100. What is the first step to solving a linear equation in one variable?

  2. What is the last step to solving a linear equation in one variable?

During Class:

☐   **Write your class notes.** Neatly write down **all** examples shown as well as key terms or phrases with definitions. If not applicable or if you were absent, watch the Lecture Series (DVD) for this section and do the same (write down the examples shown as well as key terms or phrases). Insert more paper as needed.

Class Notes/Examples	Your Notes

Answers: **1)** Multiply both sides by the LCD to clear fractions if they occur.   **2)** Check the solution by substituting it into the original equation.

Section 2.4 Solving Linear Equations

Class Notes (continued)	**Your Notes**

(Insert additional paper as needed.)

Copyright © 2013 Pearson Education, Inc.

Section 2.4 Solving Linear Equations

Practice:

☐   Complete the Vocabulary, Readiness & Video Check on page 106.

☐   Next, complete any incomplete exercises below. Check and correct your work using the answers and references at the end of this section.

Review this example:	**Your turn:**
1. Solve: $4(2x-3)+7 = 3x+5$	**2.** Solve: $5(2x-1)-2(3x)=1$

There are no fractions, so we begin with Step 2.

$$4(2x-3)+7 = 3x+5$$

Step 2. $8x-12+7 = 3x+5$

Step 3. $8x-5 = 3x+5$

Step 4. $8x-5-3x = 3x+5-3x$

$$5x-5 = 5$$
$$5x-5+5 = 5+5$$
$$5x = 10$$

Step 5. $\dfrac{5x}{5} = \dfrac{10}{5}$

$$x = 2$$

Step 6. Check: $4(2x-3)+7 = 3x+5$

$$4[2(2)-3]+7 \overset{?}{=} 3(2)+5$$
$$4(4-3)+7 \overset{?}{=} 6+5$$
$$4(1)+7 \overset{?}{=} 11$$
$$4+7 \overset{?}{=} 11$$
$$11 = 11$$

The solution is $\boxed{2.}$

Section 2.4 Solving Linear Equations

Review this example:

3. Solve: $\dfrac{x}{2} - 1 = \dfrac{2}{3}x - 3$

Your turn:

4. Solve: $\dfrac{x}{2} - 1 = \dfrac{x}{5} + 2$

We begin by clearing fractions. To do this, we multiply both sides of the equation by the LCD of 2 and 3, which is 6.

$$\frac{x}{2} - 1 = \frac{2}{3}x - 3$$

Step 1. $6\left(\dfrac{x}{2} - 1\right) = 6\left(\dfrac{2}{3}x - 3\right)$

Step 2. $6\left(\dfrac{x}{2}\right) - 6(1) = 6\left(\dfrac{2}{3}x\right) - 6(3)$

$$3x - 6 = 4x - 18$$

There are no grouping symbols and no like terms on either side of the equation, so continue with Step 4.

Step 4. $\quad 3x - 6 - 3x = 4x - 18 - 3x$

$$-6 = x - 18$$

$$-6 + 18 = x - 18 + 18$$

$$12 = x$$

Step 5. The variable is now alone, so there is no need to apply the multiplication property of equality.

Step 6. Check: $\quad \dfrac{x}{2} - 1 = \dfrac{2}{3}x - 3$

$$\dfrac{12}{2} - 1 \overset{?}{=} \dfrac{2}{3} \cdot 12 - 3$$

$$6 - 1 \overset{?}{=} 8 - 3$$

$$5 = 5$$

The solution is $\boxed{12.}$

Review this example:

5. Solve: $0.25x + 0.10(x-3) = 0.05(22)$

First clear the equation of decimals by multiplying both sides of the equation by 100.

Step 1. $0.25x + 0.10(x-3) = 0.05(22)$

                $25x + 10(x-3) = 5(22)$

Step 2. $25x + 10x - 30 = 110$

Step 3. $35x - 30 = 110$

Step 4. $35x - 30 + 30 = 110 + 30$

                    $35x = 140$

Step 5. $\dfrac{35x}{35} = \dfrac{140}{35}$

                    $x = 4$

Step 6. To check, replace x with 4 in the original equation. The solution is $\boxed{4.}$

Your turn:

6. Solve: $0.50x + 0.15(70) = 35.5$

Review this example:

7. Solve: $-2(x-5) + 10 = -3(x+2) + x$

$-2(x-5) + 10 = -3(x+2) + x$

$-2x + 10 + 10 = -3x - 6 + x$

$-2x + 20 = -2x - 6$

$-2x + 20 + 2x = -2x - 6 + 2x$

$20 = -6$

The final equation contains no variable terms, and there is no value for x that makes $20 = -6$ a true equation.

We conclude that there is $\boxed{\text{no solution}}$ to this equation.

Your turn:

8. Solve: $2(x+3) - 5 = 5x - 3(1+x)$

Section 2.4 Solving Linear Equations

	Answer	Text Ref	Video Ref			Answer	Text Ref	Video Ref
1	2	Ex 1, p. 101			**5**	4	Ex 5, pp. 103–104	
2	$\dfrac{3}{2}$		Sec 2.4, 1/5		**6**	50		Sec 2.4, 3/5
3	12	Ex 3, p. 102			**7**	no solution	Ex 6, p. 104	
4	10		Sec 2.4, 2/5		**8**	no solution		Sec 2.4, 5/5

☐   **Next, insert your homework.** Make sure you attempt all exercises asked of you and show all work, as in the exercises above. Check your answers if possible. Clearly mark any exercises you were unable to correctly complete so that you may ask questions later. DO NOT ERASE YOUR INCORRECT WORK. THIS IS HOW WE UNDERSTAND AND EXPLAIN TO YOU YOUR ERRORS.

Copyright © 2013 Pearson Education, Inc.

Section 2.5 An Introduction to Problem Solving

Before Class:

☐ Read the objectives on page 109.

☐ Read the **Helpful Hint** boxes on pages 109, 110, and 111.

☐ Complete the exercises:

1. Read the General Strategy for Problem Solving box on page 109. What are two ways to become comfortable with a problem?

2. When checking a solution, to where should you go back?

3. Once you have checked all proposed solutions, what is the last thing to do in solving a problem?

During Class:

☐ **Write your class notes.** Neatly write down **all** examples shown as well as key terms or phrases with definitions. If not applicable or if you were absent, watch the Lecture Series (DVD) for this section and do the same (write down the examples shown as well as key terms or phrases). Insert more paper as needed.

Class Notes/Examples	Your Notes

Answers and References: **1)** Answers may vary, see p. 109.   **2)** the original stated problem
3) State your conclusion.

Section 2.5 An Introduction to Problem Solving

Class Notes (continued)	**Your Notes**

(Insert additional paper as needed.)

Section 2.5 An Introduction to Problem Solving

Practice:

☐   Complete the Vocabulary, Readiness & Video Check on page 114.

☐   Next, complete any incomplete exercises below. Check and correct your work using the answers and references at the end of this section.

Review this example:

1. Twice the sum of a number and 4 is the same as four times the number, decreased by 12. Find the number.

UNDERSTAND. Read and reread the problem. If we let x = the unknown number, then "the sum of a number and 4" translates to "$x + 4$" and "four times the number" translates to "$4x$."

TRANSLATE.

twice	sum of a number and 4	is the same as	four times the number	decreased by	12
↓	↓	↓	↓	↓	↓
2	$(x+4)$	=	$4x$	−	12

SOLVE.

$$2(x+4) = 4x - 12$$
$$2x + 8 = 4x - 12$$
$$2x + 8 - 4x = 4x - 12 - 4x$$
$$-2x + 8 = -12$$
$$-2x + 8 - 8 = -12 - 8$$
$$-2x = -20$$
$$\frac{-2x}{-2} = \frac{-20}{-2}$$
$$x = 10$$

INTERPRET.

Check: Check this solution in the problem as it was originally stated. To do so, replace "number" with 10. Twice the sum of "10" and 4 is 28, which is the same as 4 times "10" decreased by 12.

State: The number is ⃝10.

Your turn:

2. Write the following as an equation, then solve.

Twice the difference of a number and 8 is equal to three times the sum of the number and 3. Find the number.

Section 2.5 An Introduction to Problem Solving

Review this example:

3. If the two walls of the Vietnam Veterans Memorial in Washington, D.C., were connected, an isosceles triangle would be formed. The measure of the third angle is 97.5° more than the measure of either of the other two angles. Find the measure of the third angle. (*Source*: National Park Service)

UNDERSTAND. Read and reread the problem. Draw a diagram and let

x = degree measure of one angle

x = degree measure of the second angle

$x + 97.5$ = degree measure of the third angle

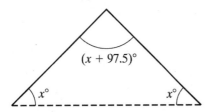

TRANSLATE. Recall that the sum of the measures of the angles of a triangle equals 180.

$$x + x + (x + 97.5) = 180$$

SOLVE.

$$x + x + (x + 97.5) = 180$$
$$3x + 97.5 = 180$$
$$3x + 97.5 - 97.5 = 180 - 97.5$$
$$3x = 82.5$$
$$\frac{3x}{3} = \frac{82.5}{3}$$
$$x = 27.5$$

INTERPRET.

Check: If $x = 27.5$, then the measure of the third angle is $x + 97.5 = 125$. The sum of the angles is then $27.5 + 27.5 + 125 = 180$, the correct sum.

State: The third angle measures $\boxed{125°}$ *

*The two walls actually meet at an angle of 125 degrees 12 minutes. The measurement of 97.5° given in the problem is an approximation.

Your turn:

4. Two angles are supplementary if their sum is 180°. The larger angle measures eight degrees more than three times the measure of a smaller angle. If x represents the measure of the smaller angle and these two angles are supplementary, find the measure of each angle.

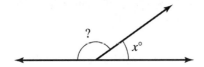

Section 2.5 An Introduction to Problem Solving

Review this example:

5. Some states have a single area code for the entire state. Two such states have area codes that are consecutive odd integers. If the sum of these integers is 1208, find the two area codes. (*Source*: North American Numbering Plan Administration)

UNDERSTAND. Read and reread the problem. Let x = the first odd integer, and $x + 2$ = the next odd integer

TRANSLATE.

$$x + (x + 2) = 1208$$

SOLVE.

$$x + (x + 2) = 1208$$
$$2x + 2 = 1208$$
$$2x + 2 - 2 = 1208 - 2$$
$$2x = 1206$$
$$\frac{2x}{2} = \frac{1206}{2}$$
$$x = 603$$

INTERPRET.

Check: If $x = 603$, then the next odd integer is $x + 2 = 603 + 2 = 605$. Notice their sum, $603 + 605 = 1208$, as needed.

State: The area codes are 603 and 605.

Your turn:

6. The measures of the angles of a triangle are 3 consecutive even integers. Find the measure of each angle.

Section 2.5 An Introduction to Problem Solving

	Answer	Text Ref	Video Ref		Answer	Text Ref	Video Ref
1	10	Ex 2, p. 110		**4**	43°, 137°		Sec 2.5, 3/4
2	$2(x-8)=3(x+3)$; -25		Sec 2.5, 1/4	**5**	603, 605	Ex 6, p. 114	
3	125°	Ex 5, pp112–113		**6**	58°, 60°, 62°		Sec 2.5, 4/4

☐　**Next, insert your homework.** Make sure you attempt all exercises asked of you and show all work, as in the exercises above. Check your answers if possible. Clearly mark any exercises you were unable to correctly complete so that you may ask questions later. DO NOT ERASE YOUR INCORRECT WORK. THIS IS HOW WE UNDERSTAND AND EXPLAIN TO YOU YOUR ERRORS.

Section 2.6 Formulas and Problem Solving

Before Class:

☐   Read the objectives on page 120.

☐   Read the **Helpful Hint** boxes on pages 121 and 126.

☐   Complete the exercises:

    1.   Write the formula that relates distance, rate, and time.

    2.   Write the formula for the perimeter of a rectangle.

    3.   Write the formula for converting degrees Celsius to degrees Fahrenheit.

During Class:

☐   **Write your class notes.** Neatly write down **all** examples shown as well as key terms or phrases with definitions. If not applicable or if you were absent, watch the Lecture Series (DVD) for this section and do the same (write down the examples shown as well as key terms or phrases). Insert more paper as needed.

Class Notes/Examples	Your Notes

Answers: **1)** $d = rt$ **2)** $P = 2l + 2w$ **3)** $F = \dfrac{9}{5}C + 32$ or $F = 1.8C + 32$

Section 2.6 Formulas and Problem Solving

Class Notes (continued)	**Your Notes**

(Insert additional paper as needed.)

　　　　　　　　Copyright © 2013 Pearson Education, Inc.

Practice:

☐ Complete any incomplete exercises below. Check and correct your work using the answers and references at the end of this section.

Review this example:

1. Portage Glacier in Alaska is about 6 miles, or 31,680 feet, long and moves 400 feet per year. Icebergs are created when the front end of the glacier flows into Portage Lake. How long does it take for ice at the head (beginning) of the glacier to reach the lake?

UNDERSTAND. Read and reread the problem. The formula needed to solve this problem is the distance formula, $d = rt$. Let
t = the time in years for ice to reach the lake
r = rate or speed of ice
d = distance from beginning of glacier to lake

TRANSLATE.

Let distance d = 31,680 feet and rate r = 400 feet per year.

$$d = r \cdot t$$
$$31,680 = 400 \cdot t$$

SOLVE.

Solve the equation for t.
$$\frac{31,680}{400} = \frac{400 \cdot t}{400}$$
$$79.2 = t$$

INTERPRET.

Check: Substitute 79.2 for t and 400 for r in the distance formula and check to see that the distance is 31,680 feet.

State: It takes ⟨79.2 years⟩ for the ice at the head of Portage Glacier to reach the lake.

Your turn:

2. The Cat is a high-speed catamaran auto ferry that operates between Bar Harbor, Maine, and Yarmouth, Nova Scotia. The Cat can make the 138-mile trip in about $2\frac{1}{2}$ hours. Find the catamaran speed for this trip. (*Source*: Bay Ferries)

Section 2.6 Formulas and Problem Solving

Review this example:

3. Charles Pecot can afford enough fencing to enclose a rectangular garden with a perimeter of 140 feet. If the width of his garden must be 30 feet, find the length.

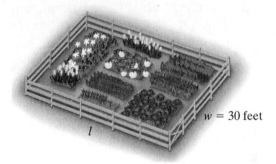

$w = 30$ feet

l

Your turn:

4. An architect designs a rectangular flower garden such that the width is exactly two-thirds of the length. If 260 feet of antique picket fencing are to be used to enclose the garden, find the dimensions of the garden.

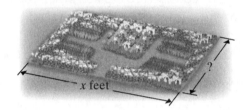

x feet

?

UNDERSTAND. Read and reread the problem. The formula needed to solve this problem is the formula for the perimeter of a rectangle, $P = 2l + 2w$, where

l = the length of the rectangular garden
w = the width of the rectangular garden
P = the perimeter of the garden

TRANSLATE.

Let perimeter $P = 140$ feet and width $w = 30$ feet.
$$P = 2l + 2w$$
$$140 = 2l + 2(30)$$

SOLVE.

$$140 = 2l + 2(30)$$
$$140 = 2l + 60$$
$$140 - 60 = 2l + 60 - 60$$
$$80 = 2l$$
$$40 = l$$

INTERPRET.

Check: Substitute 40 for l and 30 for w in the perimeter formula and check to see that the perimeter is 140 feet.

State: The length of the garden is 40 feet.

Section 2.6 Formulas and Problem Solving

Review this example:

5. The average minimum temperature for July in Shanghai, China, is 77° Fahrenheit. Find the equivalent temperature in degrees Celsius.

UNDERSTAND. Read and reread the problem. A formula that can be used to solve this problem is the formula for converting degrees Celsius to degrees Fahrenheit, $F = \dfrac{9}{5}C + 32$, where

C = temperature in degrees Celsius, and
F = temperature in degrees Fahrenheit.

TRANSLATE. Let degrees Fahrenheit $F = 77$.

$$F = \frac{9}{5}C + 32$$

$$77 = \frac{9}{5}C + 32$$

SOLVE.

$$77 = \frac{9}{5}C + 32$$

$$77 - 32 = \frac{9}{5}C + 32 - 32$$

$$45 = \frac{9}{5}C$$

$$\frac{5}{9} \cdot 45 = \frac{5}{9} \cdot \frac{9}{5}C$$

$$25 = C$$

INTERPRET.

Check: Replace C with 25 and F with 77 in the formula and see that a true statement results.

State: 77° Fahrenheit is equivalent to 25° Celsius.

Your turn:

6. Convert Nome, Alaska's 14°F high temperature to Celsius.

Section 2.6 Formulas and Problem Solving

Review this example:

7. Solve $P = 2l + 2w$ for w.

This formula relates the perimeter of a rectangle to its length and width. Find the term containing the variable w. Get this term, $2w$, alone.

$$P = 2l + 2w$$

$$P - 2l = 2l + 2w - 2l$$

$$P - 2l = 2w$$

$$\frac{P - 2l}{2} = \frac{2w}{2}$$

$$\frac{P - 2l}{2} = w$$

Your turn:

8. Solve the formula $S = 2\pi rh + 2\pi r^2$ for h.

	Answer	Text Ref	Video Ref		Answer	Text Ref	Video Ref
1	79.2 yr	Ex 1, pp. 120–121		5	$25°\,C$	Ex 3 pp. 122–123	
2	55.2 mph		Sec 2.6, 2/5	6	$-10°\,C$		Sec 2.6, 1/5
3	40 ft	Ex 2, p. 120		7	$w = \dfrac{P - 2l}{2}$	Ex 7, pp.125–126	
4	length: 78 ft, width: 52 ft		Sec 2.6, 3/5	8	$h = \dfrac{S - 2\pi r^2}{2\pi r}$		Sec 2.6, 5/5

☐ **Next, insert your homework.** Make sure you attempt all exercises asked of you and show all work, as in the exercises above. Check your answers if possible. Clearly mark any exercises you were unable to correctly complete so that you may ask questions later. DO NOT ERASE YOUR INCORRECT WORK. THIS IS HOW WE UNDERSTAND AND EXPLAIN TO YOU YOUR ERRORS.

Section 2.7 Percent and Mixture Problem Solving

Before Class:

☐   Read the objectives on page 131.

☐   Read the **Helpful Hint** boxes on pages 133 and 134.

☐   Complete the exercises:

1. The percents in a circle graph should have a sum of _____ .

2. Discounts are _____ the original price while markups are

    _____ the original price.

3. Percent increase or percent decrease is a common way to describe how some

    measurement has_____ .

During Class:

☐   **Write your class notes.** Neatly write down **all** examples shown as well as key terms or phrases with definitions. If not applicable or if you were absent, watch the Lecture Series (DVD) for this section and do the same (write down the examples shown as well as key terms or phrases). Insert more paper as needed.

Class Notes/Examples	**Your Notes**

Answers: **1)** 100%   **2)** subtracted from, added to   **3)** increased or decreased

Section 2.7 Percent and Mixture Problem Solving

Class Notes (continued)	Your Notes

(Insert additional paper as needed.)

Section 2.7 Percent and Mixture Problem Solving

Practice:

☐ Complete the Vocabulary, Readiness & Video Check on page 138.

☐ Next, complete any incomplete exercises below. Check and correct your work using the answers and references at the end of this section.

Review this example: **1.** The number 63 is what percent of 72?	**Your turn:** **2.** Find 23% of 20.

Review this example:

1. The number 63 is what percent of 72?

UNDERSTAND. Read and reread the problem.
Let x = the unknown percent.

TRANSLATE.

the number 63 is what percent of 72
  ↓   ↓   ↓   ↓ ↓
  63   = x · 72

SOLVE.

$$63 = 72x$$
$$0.875 = x$$
$$87.5\% = x$$

INTERPRET.
Check: Verify that 87.5% of 72 is 63.
State: The number 63 is 87.5% of 72.

Your turn:

2. Find 23% of 20.

Review this example:

3. The number 120 is 15% of what number?

UNDERSTAND. Read and reread the problem.
Let x = the unknown number.

TRANSLATE.

the number 120 is 15% of what number
  ↓   ↓ ↓ ↓   ↓
  120   = 15% · x

SOLVE. $120 = 0.15x$
     $800 = x$

INTERPRET.
Check: Check the proposed solution by finding 15% of 800 and verifying that the result is 120.

State: 120 is 15% of 800.

Your turn:

4. The number 45 is 25% of what number?

Section 2.7 Percent and Mixture Problem Solving

Review this example:

5. Cell Phones Unlimited recently reduced the price of a $140 phone by 20%. What is the discount and the new price?

UNDERSTAND. Read and reread the problem.
discount = percent · original price
new price = original price – discount

TRANSLATE and SOLVE.
discount = percent · original price
      = 20% · $140 = 0.20 · $140 = $28
new price = original price – discount
      = $140 – $28 = $112

INTERPRET.
Check: Check your calculations in the formulas, and also see if the results are reasonable.

State: The discount in price is $28 and the new price is $112.

Your turn:

6. A birthday celebration meal is $40.50 including tax. Find the total cost if a 15% tip is added to the cost. If needed, round to the nearest cent.

Review this example:

7. The cost of public college rose from $4020 in 1996 to $7610 in 2011. Find the percent increase, rounded to the nearest tenth of a percent. (*Source*: The College Board)

UNDERSTAND. Read and reread the problem.
Let x = the percent increase.

TRANSLATE.
increase = new cost – old cost
increase = $7610 – $4020 = $3590
The percent increase is always a percent of the original number (the old cost).
increase = percent increase (x) · old cost
$3590 = x$ · $4020

SOLVE.
$$3590 = x \cdot 4020$$
$$0.893 \approx x$$
$$89.3\% \approx x$$

INTERPRET.
Check: Check the proposed solution by finding the increase in cost (89.3% of $4020). Then add this to $4020 to find the new cost (about $7610).
State: The percent increase in cost is about 89.3%

Your turn:

8. Iceberg lettuce is grown and shipped to stores for about 40 cents a head, and consumers purchase it for about 86 cents a head. Find the percent increase. (*Source: Statistical Abstract of the United States*)

Section 2.7 Percent and Mixture Problem Solving

Review this example:

9. A chemist needs 12 liters of a 50% acid solution. The stockroom has only 40% and 70% solutions. How much of each solution should be mixed together to form 12 liters of a 50% solution?

UNDERSTAND. Read and reread the problem. Let x = number of liters of 40% solution; then $12 - x$ = number of liters of 70% solution.

TRANSLATE. The table summarizes the information given.

	Number · of Liters	Acid Strength	= Amount of Acid
40% Solution	x	40%	0.40x
70% Solution	$12 - x$	70%	0.70(12 − x)
50% Solution Needed	12	50%	0.50(12)

The amount of acid in the final solution is the sum of the acid amounts in the two beginning solutions.
$$0.40x + 0.70(12 - x) = 0.50(12)$$

SOLVE.
$$0.40x + 0.70(12 - x) = 0.50(12)$$
$$0.4x + 8.4 - 0.7x = 6$$
$$-0.3x + 8.4 = 6$$
$$-0.3x = -2.4$$
$$x = 8$$

INTERPRET.
Check: In 8 liters of a 40% acid solution there are $0.40(8) = 3.2$ liters of acid. In $12 - 8 = 4$ liters of a 70% acid solution there are $0.70(4) = 2.8$ liters of acid. In 12 liters of 50% acid solution there are $0.50(12) = 6$ liters of acid. $3.2 + 2.8 = 6$, so the amounts are correct.

State: If 8 liters of the 40% solution are mixed with 4 liters of the 70% solution, the result is 12 liters of a 50% solution.

Your turn:

10. How much of an alloy that is 20% copper should be mixed with 200 ounces of an alloy that is 50% copper in order to get an alloy that is 30% copper?

Alloy	Ounces ·	Copper strength	= Amount of copper
20%			
50%	200		
30%			

Section 2.7 Percent and Mixture Problem Solving

	Answer	Text Ref	Video Ref			Answer	Text Ref	Video Ref
1	87.5%	Ex 1, p.132			6	$46.58		Sec 2.7, 3/6
2	4.6		Sec 2.7, 1/6		7	89.3%	Ex 5, p. 135	
3	800	Ex 2, pp. 132–133			8	115% increase		Sec 2.7, 4/6
4	180		Sec 2.7, 2/6		9	8 liters of 40% solution, 4 liters of 70% solution	Ex 7, pp. 136-137	
5	discount: $28, new price: $112	Ex 4, p.134			10	400 ounces of 20% alloy, 600 ounces of 30% alloy		Sec 2.7, 6/6

☐ **Next, insert your homework.** Make sure you attempt all exercises asked of you and show all work, as in the exercises above. Check your answers if possible. Clearly mark any exercises you were unable to correctly complete so that you may ask questions later. DO NOT ERASE YOUR INCORRECT WORK. THIS IS HOW WE UNDERSTAND AND EXPLAIN TO YOU YOUR ERRORS.

Section 2.8 Further Problem Solving

Before Class:

☐   Read the objectives on page 143.

☐   Complete the exercises:

     1.   If one car travels at a rate of x miles per hour and a second car travels 10 miles per hour faster, write an expression for the rate of the second car.

     2.   Write an expression and solve for the number of $5 bills that equal $40.

During Class:

☐   **Write your class notes.** Neatly write down **all** examples shown as well as key terms or phrases with definitions. If not applicable or if you were absent, watch the Lecture Series (DVD) for this section and do the same (write down the examples shown as well as key terms or phrases). Insert more paper as needed.

Class Notes/Examples	**Your Notes**

Answers: **1)** $x + 10$ **2)** $x = \dfrac{\$40}{\$5} = 8$ bills

Section 2.8 Further Problem Solving

Class Notes (continued)	**Your Notes**

(Insert additional paper as needed.)

Section 2.8 Further Problem Solving

Practice:

☐   Complete the Vocabulary, Readiness & Video Check on page 138.

☐   Next, complete any incomplete exercises below. Check and correct your work using the answers and references at the end of this section.

Review this example:

1. Suppose two trains leave Neosho, Missouri, at the same time. One travels north and the other travels south at a speed that is 15 miles per hour faster. In 2 hours, the trains are 230 miles apart. Find the speed of each train.

UNDERSTAND. Read and reread the problem. Let x = speed of train traveling north; $x + 15$ = speed of train traveling south.

TRANSLATE. The table summarizes the information given.

	r	$\cdot$	t	$=$	d
North train	x		2		$2x$
South train	$x + 15$		2		$2(x + 15)$

The total distance between the trains is 230 miles.
$2x + 2(x + 15) = 230$

SOLVE. $2x + 2(x + 15) = 230$

$2x + 2x + 30 = 230$

$4x + 30 = 230$

$4x = 200$

$\dfrac{4x}{4} = \dfrac{200}{4}$

$x = 50$

INTERPRET.
Check: Recall that x is the speed of the train traveling north, or 50 mph. In 2 hours, this train travels 2(50) = 100 miles. The speed of the train traveling south is $x + 15$ or 50 + 15 = 65 mph. In 2 hours, this train travels 2(65) = 130 miles. The total distance of the trains is 100 miles + 130 miles = 230 miles, the required distance.

State: The northbound train's speed is 50 mph.
             The southbound train's speed is 65 mph.

Your turn:

2. Two hikers are 11 miles apart and walking toward each other. They meet in 2 hours. Find the rate of each hiker if one hiker walks 1.1 miles per hour faster than the other.

	r	$\cdot$	t	$=$	d
Hiker					
Other hiker					

Section 2.8 Further Problem Solving

Review this example:

3. Part of the proceeds from a local talent show was $2420 worth of $10 and $20 bills. If there were 37 more $20 bills than $10 bills, find the number of each denomination.

UNDERSTAND. Read and reread the problem. Let x = number of $10 bills and $x + 37$ = number of $20 bills.

TRANSLATE.

Denomination	Number of Bills	Value of Bills (in dollars)
$10 bills	x	$10x$
$20 bills	$x + 37$	$20(x + 37)$

Since the total value of these bills is $2420,
$$10x + 20(x + 37) = 2420$$

SOLVE.
$$10x + 20(x + 37) = 2420$$
$$10x + 20x + 740 = 2420$$
$$30x + 740 = 2420$$
$$30x = 1680$$
$$\frac{30x}{30} = \frac{1680}{30}$$
$$x = 56$$

INTERPRET.
Check: Since x represents the number of $10 bills, we have 56 $10 bills and 56 + 37 = 93 $20 bills. The total amount is $10(56) + $20(93) = $2420, the correct total.

State: There are 56 $10 bills and 93 $20 bills.

Your turn:

4. Part of the proceeds from a garage sale was $280 worth of $5 and $10 bills. If there were 20 more $5 bills than $10 bills, find the number of each denomination.

	Number of Bills	Value of Bills
$5 bills		
$10 bills		
Total		

	Answer	Text Ref	Video Ref		Answer	Text Ref	Video Ref
1	50 mph north 65 mph south	Ex 2, pp. 144–145		**3**	56 $10 bills, 93 $20 bills	Ex 3 pp. 146–146	
2	2.2 mph, 3.3 mph		Sec 2.8, 2/4	**4**	12 $10 bills, 32 $5 bills		Sec 2.8, 3/4

☐   **Next, insert your homework.** Make sure you attempt all exercises asked of you and show all work, as in the exercises above. Check your answers if possible. Clearly mark any exercises you were unable to correctly complete so that you may ask questions later. DO NOT ERASE YOUR INCORRECT WORK. THIS IS HOW WE UNDERSTAND AND EXPLAIN TO YOU YOUR ERRORS.

Section 2.9 Solving Linear Inequalities

Before Class:

☐   Read the objectives on page 150.

☐   Read the **Helpful Hint** boxes on pages 151, 152, 153, and 154.

☐   Complete the exercises:

1. A linear inequality in one variable can be written in the form _____,
    where a, b, and c are _____ numbers and a is not _____?

2. For real numbers a, b, and c, if $a < b$ and $ac > bc$, then c is _____.

3. What are inequalities containing two inequality symbols called?

During Class:

☐   **Write your class notes.** Neatly write down **all** examples shown as well as key terms or
    phrases with definitions. If not applicable or if you were absent, watch the Lecture Series
    (DVD) for this section and do the same (write down the examples shown as well as key terms
    or phrases). Insert more paper as needed.

Class Notes/Examples	Your Notes

Answers: **1)** $ax + b < c$, real, 0   **2)** negative   **3)** compound inequalities

Section 2.9 Solving Linear Inequalities

Class Notes (continued)

Your Notes

(Insert additional paper as needed.)

Section 2.9 Solving Linear Inequalities

Practice:

☐   Complete the Vocabulary, Readiness & Video Check on page 158.

☐   Next, complete any incomplete exercises below. Check and correct your work using the answers and references at the end of this section.

Review this example:	**Your turn:**
1. Graph $x \geq -1$. Then write the solution set in interval notation.	2. Graph $x \leq -1$. Then write the solution set in interval notation.

We place a bracket -1 since the inequality symbol is $\geq$ and -1 is greater than or equal to -1. Then we shade to the right of -1.

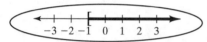

In interval notation, this is $[-1, \infty)$

Review this example:	**Your turn:**
3. Solve $x + 4 \leq -6$ for x. Graph the solution set and write it in interval notation.	4. Solve $x - 2 \geq -7$. Graph the solution set and write it in interval notation.

$$x + 4 \leq -6$$
$$x + 4 - 4 \leq -6 - 4$$
$$x \leq -10$$

The solution set $(-\infty, -10]$ is graphed as shown.

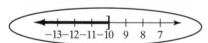

Review this example:	**Your turn:**
5. Solve $-2x \leq -4$. Graph the solution set and write it in interval notation.	6. Solve $-8x \leq -16$. Graph the solution set and write it in interval notation.

$$-2x \leq -4$$
$$\frac{-2x}{-2} \geq \frac{-4}{-2}$$
$$x \geq 2$$

The solution set $[2, \infty)$ is graphed as shown.

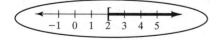

Section 2.9 Solving Linear Inequalities

Review this example:	Your turn:
7. Solve $2(x-3)-5 \leq 3(x+2)-18$. Graph the solution set and write it in interval notation.	**8.** Solve $2(x-4)-3x \leq -(4x+1)+2x$. Graph the solution set and write it in interval notation.

$$2(x-3)-5 \leq 3(x+2)-18$$
$$2x-6-5 \leq 3x+6-18$$
$$2x-11 \leq 3x-12$$
$$-x-11 \leq -12$$
$$-x \leq -1$$
$$\frac{-x}{-1} \geq \frac{-1}{-1}$$
$$x \geq 1$$

The graph of the solution set $[1, \infty)$ is shown.

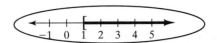

Review this example:	Your turn:
9. Solve $-1 \leq 2x-3 < 5$. Graph the solution set and write it in interval notation.	**10.** Solve $-6 < 3(x-2) \leq 8$. Graph the solution set and write it in interval notation.

$$-1 \leq 2x-3 < 5$$
$$-1+3 \leq 2x-3+3 < 5+3$$
$$2 \leq 2x < 8$$
$$\frac{2}{2} \leq \frac{2x}{2} < \frac{8}{2}$$
$$1 \leq x < 4$$

The graph of the solution set $[1, 4)$ is shown.

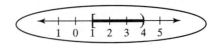

Review this example:

11. A couple may spend at most $2000 for a reception. The reception hall charges a $100 cleanup fee plus $36 per person. Find the greatest number of people that they can invite and still stay within the budget.

UNDERSTAND. Read and reread the problem. Let x = the number of people who attend the reception.

TRANSLATE.

cleanup fee	+	cost/person	must be less than or equal to	$2000
↓	↓	↓	↓	↓
100	+	36x	≤	2000

SOLVE. $100 + 36x \leq 2000$

$$36x \leq 1900$$

$$x \leq 52\frac{7}{9}$$

INTERPRET.
Check: Round down to the nearest whole, or 52. Notice that if 52 people attend, the cost is $1972. If 53 attend, the cost is $2008, which is more than $2000.

State: The most people that can be invited is $\boxed{52.}$

Your turn:

12. Find the values for x so that the perimeter of this rectangle is no greater than 100 centimeters.

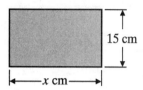

15 cm

x cm

Section 2.9 Solving Linear Inequalities

	Answer	Text Ref	Video Ref		Answer	Text Ref	Video Ref
1	$[-1, \infty)$	Ex 1, p. 151		7	$[1, \infty)$	Ex 8, pp. 155–156	
2	$(-\infty, -1]$		Sec 2.9, 1/9	8	$(3, \infty)$		Sec 2.9, 6/9
3	$(-\infty, -10]$	Ex 2, p. 152		9	$[1, 4)$	Ex 10, p. 156	
4	$[-5, \infty)$		Sec 2.9, 3/9	10	$\left(0, \dfrac{14}{3}\right]$		Sec 2.9, 8/9
5	$[2, \infty)$	Ex 3, p. 153		11	52 people	Ex 13, p. 158	
6	$[-2, \infty)$		Sec 2.9, 4/9	12	less than or equal to 35 centimeters		Sec 2.9, 9/9

☐ **Next, insert your homework.** Make sure you attempt all exercises asked of you and show all work, as in the exercises above. Check your answers if possible. Clearly mark any exercises you were unable to correctly complete so that you may ask questions later. DO NOT ERASE YOUR INCORRECT WORK. THIS IS HOW WE UNDERSTAND AND EXPLAIN TO YOU YOUR ERRORS.

Preparing for the Chapter 2 Test

Start preparing for your Chapter 2 Test as soon as possible. Pay careful attention to any instructor discussion about this test, especially discussion on what sections you will be responsible for, etc.

☐   Work the Chapter 2 Vocabulary Check on page 162.

☐   Read your Class Notes/Examples for each section covered on your Chapter 2 Test. Look for any unresolved questions you may have.

☐   Complete as many of the Chapter 2 Review exercises as possible (page 167). Remember, the odd answers are in the back of your text.

☐   **Most important:** Place yourself in "test" conditions (see below) and work the Chapter 2 Test (page 170) as a practice test the day before your actual test. To honestly assess how you are doing, try the following:

- Work on a few blank sheets of paper.
- Give yourself the same amount of time you will be given for your actual test.
- Complete this Chapter 2 Practice Test without using your notes or your text.
- If you have any time left after completing this practice test, check your work and try to find any errors on your own.
- Once done, use the back of your book to check ALL answers.
- Try to correct any errors on your own.
- Use the Chapter Test Prep Video (CTPV) to correct any errors you were unable to correct on your own. You can find these videos in the Interactive DVD Lecture Series, in MyMathLab, and on YouTube. Search Martin-Gay Beginning Algebra and click "Channels."

I wish you the best of luck….Elayn Martin-Gay

Section 3.1 Reading Graphs and the Rectangular Coordinate System

Before Class:

☐   Read the objectives on page 174.

☐   Read the **Helpful Hint** box on page 177.

☐   Complete the exercises:

1. Does the order in which coordinates are listed matter?

2. The graph of paired data as points in the rectangular coordinate system is called a

    _____ diagram.

3. If one value of an ordered pair solution of an equation is known, how can the other value be determined?

During Class:

☐   **Write your class notes.** Neatly write down **all** examples shown as well as key terms or phrases with definitions. If not applicable or if you were absent, watch the Lecture Series (DVD) for this section and do the same (write down the examples shown as well as key terms or phrases). Insert more paper as needed.

Class Notes/Examples	**Your Notes**

Answers: **1)** yes   **2)** scatter   **3)** Replace one variable in the equation by its known value and solve for the unknown variable.

Section 3.1 Reading Graphs and the Rectangular Coordinate System

Class Notes (continued)	**Your Notes**

(Insert additional paper as needed.)

Section 3.1 Reading Graphs and the Rectangular Coordinate System

Practice:

☐   Complete the Vocabulary, Readiness & Video Check on page 183.

☐   Next, complete any incomplete exercises below. Check and correct your work using the answers and references at the end of this section.

Review this example:

1. The following bar graph shows the estimated number of Internet users worldwide by region, as of a recent year.

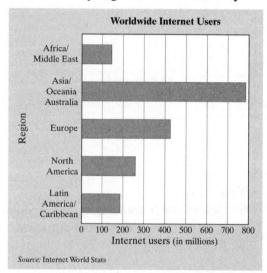

Source: Internet World Stats

    a. Find the region that has the most Internet users and approximate the number of users.

    b. How many more users are in the North America region than the Latin America/Caribbean region?

a. Look for the longest bar, which is the bar representing Asia/Oceania/Australia. Move from the right edge of this bar downward to the Internet user axis. This region has approximately 785 million Internet users.

b. The North America region has approximately 260 million Internet users. The Latin America/Caribbean region has approximately 187 million Internet users. The North America region has 260-187 = 73 million more Internet users.

Your turn:

2. The following bar graph shows the top 10 tourist destinations and the number of tourists that visit each country per year.

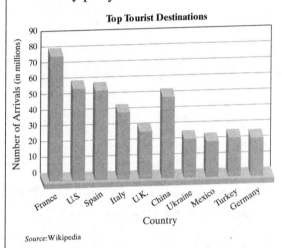

Source: Wikipedia

    a. Which country shown is the most popular tourist destination?

    b. Which countries shown have more than 40 million tourists per year?

    c. Estimate the number of tourists per year whose destination is Italy.

Section 3.1 Reading Graphs and the Rectangular Coordinate System

Review this example:

3. On a single coordinate system, plot each ordered pair.

    a. $(5,3)$ b. $(-5,3)$ c. $(-2,-4)$

    d. $(1,-2)$ e. $(0,0)$ f. $(0,2)$

    g. $(-5,0)$ h. $\left(0,-5\dfrac{1}{2}\right)$

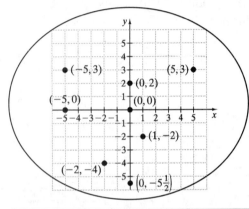

Your turn:

4. Plot each ordered pair.

    a. $(1,5)$ b. $(-5,-2)$

    c. $(-3,0)$ d $(0,-1)$

    e. $(2,-4)$ f. $\left(-1,4\dfrac{1}{2}\right)$

Review this example:

5. Determine whether each ordered pair is a solution of the equation $x-2y=6$.

    a. $(6,0)$ b. $(0,3)$

a. Let $x=6$ and $y=0$ in the equation $x-2y=6$.

$$x-2y=6$$
$$6-2(0)=6$$
$$6-0=6$$
$$6=6 \quad \text{True}$$

$(6,0)$ is a solution, since $6=6$ is a true statement.

b. Let $x=0$ and $y=3$.

$$x-2y=6$$
$$0-2(3)=6$$
$$0-6=6$$
$$-6=6 \quad \text{False}$$

$(0,3)$ is not a solution, since $-6=6$ is a false statement.

Your turn:

6. Determine whether each ordered pair is a solution of the linear equation $2x+y=7$.

$$(3,1), \ (7,0), \ (0,7)$$

Section 3.1 Reading Graphs and the Rectangular Coordinate System

Review this example:

7. Complete the following ordered pair solutions for the equation $3x + y = 12$.

   a. $(\quad , 6)$ b. $(-1, \quad)$

a. In the ordered pair $(\quad , 6)$, the y-value is 6. Let $y = 6$ in the equation and solve for x.

   $3x + y = 12$

   $3x + 6 = 12$

   $\quad 3x = 6$

   $\quad\quad x = 2$ The ordered pair is $(2, 6)$.

b. In the ordered pair $(-1, \quad)$, the x-value is -1. Let $x = -1$ in the equation and solve for y.

   $3x + y = 12$

   $3(-1) + y = 12$

   $\quad -3 + y = 12$

   $\quad\quad y = 15$ The ordered pair is $(-1, 15)$.

Your turn:

8. Complete each ordered pair so that it is a solution of the linear equation $x - 4y = 4$.

   $(\quad , -2), (4, \quad)$

Section 3.1 Reading Graphs and the Rectangular Coordinate System

	Answer	Text Ref	Video Ref		Answer	Text Ref	Video Ref
1	a. Asia/Oceania/Australia, 785 million b. 73 million	Ex 1, p. 174		5	a. yes b. no	Ex 5a, b, pp. 179–180	
2	a. France b. France, U.S., Spain, Italy, China c. 43 million	Sec 3.1, 1-3/13,		6	yes, no, yes		Sec 3.1, 12/13
3		Ex 3a–h, p. 177		7	a. (2, 6) b. (−1, 15)	Ex 6b, c, p. 180–181	
4		Sec 3.1, 4–9/13		8	(−4, −2), (4, 0)		Sec 3.1, 13/3

☐ **Next, insert your homework.** Make sure you attempt all exercises asked of you and show all work, as in the exercises above. Check your answers if possible. Clearly mark any exercises you were unable to correctly complete so that you may ask questions later. DO NOT ERASE YOUR INCORRECT WORK. THIS IS HOW WE UNDERSTAND AND EXPLAIN TO YOU YOUR ERRORS.

Section 3.2 Graphing Linear Equations

Before Class:

☐ Read the objectives on page 189.

☐ Read the **Helpful Hint** boxes on pages 189, 191, 192, and 195.

☐ Complete the exercises:

1. Write the standard form of a linear equation in two variables.

2. The graph of $y = mx + b$ crosses the y-axis at _____ .

3. A straight line is determined by _____ points, but graphing a

_____ point serves as a check.

During Class:

☐ **Write your class notes.** Neatly write down **all** examples shown as well as key terms or phrases with definitions. If not applicable or if you were absent, watch the Lecture Series (DVD) for this section and do the same (write down the examples shown as well as key terms or phrases). Insert more paper as needed.

Class Notes/Examples	**Your Notes**

Answers: **1)** $Ax + By = C$ **2)** $(0, b)$ **3)** two, third

Section 3.2 Graphing Linear Equations

Class Notes (continued)	**Your Notes**

(Insert additional paper as needed.)

Section 3.2 Graphing Linear Equations

Practice:

☐   Complete any incomplete exercises below. Check and correct your work using the answers and references at the end of this section.

Review this example:

1. Graph the linear equation $-5x + 3y = 15$.

Find three ordered pair solutions of $-5x + 3y = 15$.

Let $x = 0$.	Let $y = 0$.
$-5x + 3y = 15$	$-5x + 3y = 15$
$-5 \cdot 0 + 3y = 15$	$-5x + 3 \cdot 0 = 15$
$0 + 3y = 15$	$-5x + 0 = 15$
$3y = 15$	$-5x = 15$
$y = 5$	$x = -3$

Let $x = -2$.

$$-5x + 3y = 15$$
$$-5(-2) + 3y = 15$$
$$10 + 3y = 15$$
$$3y = 5$$
$$y = \frac{5}{3}$$

The ordered pairs are $(0, 5)$, $(-3, 0)$, and $\left(-2, \dfrac{5}{3}\right)$.

The graph is the line through the three points.

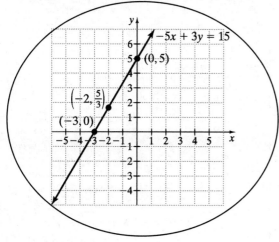

Your turn:

2. Graph the linear equation $x - 2y = 6$.

Section 3.2 Graphing Linear Equations

Review this example:

3. Graph the linear equation $y = 3x$.

Find three ordered pair solutions.
If $x = 2$, $y = 3 \cdot 2 = 6$.
If $x = 0$, $y = 3 \cdot 0 = 0$
If $x = -1$, $y = 3 \cdot -1 = -3$.
Graph the ordered pairs and draw a line through the plotted points.

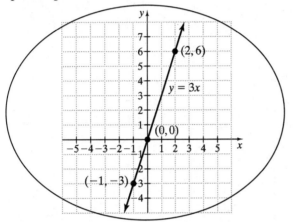

Your turn:

4. Graph the linear equation $x = -3y$.

Review this example:

5. Graph the linear equation $y = -\dfrac{1}{3}x + 2$.

Find three ordered pair solutions, graph the solutions, and draw a line through the plotted solutions. To avoid fractions, choose x values that are multiples of 3.

If $x = 6$, then $y = -\dfrac{1}{3} \cdot 6 + 2 = -2 + 2 = 0$.

If $x = 0$, then $y = -\dfrac{1}{3} \cdot 0 + 2 = 0 + 2 = 2$.

If $x = -3$, then $y = -\dfrac{1}{3} \cdot -3 + 2 = 1 + 2 = 3$.

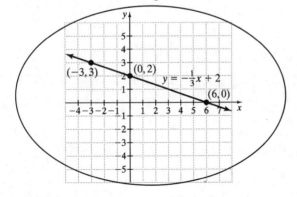

Your turn:

6. Graph the linear equation

$$y = \dfrac{1}{2}x + 2.$$

Section 3.2 Graphing Linear Equations

	Answer	Text Ref	Video Ref
1		Ex 3, p. 191	
2			Sec 3.2, 5/7
3		Ex 4, p. 192	
4			Sec 3.2, 6/7
5		Ex 5, p. 192	
6			Sec 3.2, 7/7

Section 3.2 Graphing Linear Equations

☐ **Next, insert your homework.** Make sure you attempt all exercises asked of you and show all work, as in the exercises above. Check your answers if possible. Clearly mark any exercises you were unable to correctly complete so that you may ask questions later. DO NOT ERASE YOUR INCORRECT WORK. THIS IS HOW WE UNDERSTAND AND EXPLAIN TO YOU YOUR ERRORS.

Before Class:

☐ Read the objectives on page 199.

☐ Read the **Helpful Hint** boxes on page 199.

☐ Complete the exercises:

    1. Write the general form of the equation of a vertical line.

    2. Write the general form of the equation of a horizontal line.

    3. How many x- and y-intercepts can a line have? Explain.

During Class:

☐ **Write your class notes.** Neatly write down **all** examples shown as well as key terms or phrases with definitions. If not applicable or if you were absent, watch the Lecture Series (DVD) for this section and do the same (write down the examples shown as well as key terms or phrases). Insert more paper as needed.

Class Notes/Examples	Your Notes

Answers: **1)** $x = c$ **2)** $y = c$ **3)** one intercept if the line is vertical, horizontal, or passes through the origin; otherwise two intercepts

Section 3.3 Intercepts

Class Notes (continued)	Your Notes

(Insert additional paper as needed.)

Practice:

☐   Complete the Vocabulary, Readiness & Video Check on page 204.

☐   Next, complete any incomplete exercises below. Check and correct your work using the answers and references at the end of this section.

Review this example:	**Your turn:**
1. Identify the x- and y-intercepts.	**2.** Identify the intercepts.

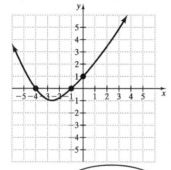 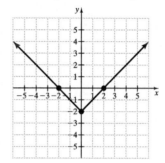

x-intercepts: $\boxed{(-4,0),(-1,0)}$ y-intercept: $\boxed{(0,1)}$

Review this example:

3. Graph $x - 3y = 6$ by finding and plotting intercepts.

Let $y = 0$.	Let $x = 0$.
$x - 3y = 6$	$x - 3y = 6$
$x - 3(0) = 6$	$0 - 3y = 6$
$x - 0 = 6$	$-3y = 6$
$x = 6$	$y = -2$

The x-intercept is $(6,0)$ and the y-intercept is $(0,-2)$. Find a third ordered pair to check. If $y = -1$, then $x = 3$. Plot the points.

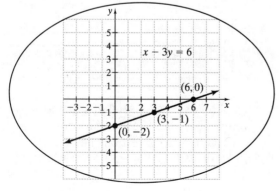

Your turn:

4. Graph the linear equation $-x + 2y = 6$ by finding and plotting its intercepts.

Section 3.3 Intercepts

Review this example:

5. Graph $y = -3$.

For any x-value chosen, y is -3. If we choose 4, 1, and -2 as x-values, the ordered pair solutions are $(4,-3)$, $(1,-3)$, and $(-2,-3)$. The graph is a horizontal line with y-intercept $(0,-3)$ and no x-intercept.

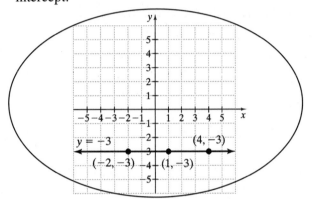

Your turn:

6. Graph $x = -3$.

	Answer	Text Ref	Video Ref		Answer	Text Ref	Video Ref
1	$(-4,0),(-1,0);(0,1)$	Ex 2, p. 199		4			Sec 3.3, 3/6
2	$(-2,0),(2,0);(0,-2)$		Sec 3.3, 2/6	5	See graph above.	Ex 10, p. 203	
3		Ex 6, pp. 200–201		6			Sec 3.3, 6/6

☐ **Next, insert your homework.** Make sure you attempt all exercises asked of you and show all work, as in the exercises above. Check your answers if possible. Clearly mark any exercises you were unable to correctly complete so that you may ask questions later. DO NOT ERASE YOUR INCORRECT WORK. THIS IS HOW WE UNDERSTAND AND EXPLAIN TO YOU YOUR ERRORS.

Section 3.4 Slope and Rate of Change

Before Class:

☐   Read the objectives on page 207.

☐   Read the **Helpful Hint** boxes on pages 207, 208, 209, 211, and 213.

☐   Complete the exercises:

1. Does the slope of a line depend on which two points on the line are used in the calculation of the slope?

2. To decide whether a line "goes up" or "goes down," always follow the line from

_____ to _____ .

3. Write the general slope-intercept form of the equation of a line.

4. Two lines that lie in the same plane and meet at a 90° angle are

_____ .

During Class:

☐   **Write your class notes.** Neatly write down **all** examples shown as well as key terms or phrases with definitions. If not applicable or if you were absent, watch the Lecture Series (DVD) for this section and do the same (write down the examples shown as well as key terms or phrases). Insert more paper as needed.

Class Notes/Examples	Your Notes

Answers: **1)** no   **2)** left, right   **3)** $y = mx + b$ **4)** perpendicular

Section 3.4 Slope and Rate of Change

Class Notes (continued)	**Your Notes**

(Insert additional paper as needed.)

Section 3.4 Slope and Rate of Change

Practice:

☐   Complete the Vocabulary, Readiness & Video Check on pages 216–217.

☐   Next, complete any incomplete exercises below. Check and correct your work using the answers and references at the end of this section.

Review this example:

1. Find the slope of the line through $(-1, 5)$ and $(2, -3)$. Graph the line.

If we let (x_1, y_1) be $(-1, 5)$, and (x_2, y_2) be $(2, -3)$, then, by the definition of slope,

$$m = \frac{y_2 - y_1}{x_2 - x_1}$$

$$= \frac{-3 - 5}{2 - (-1)}$$

$$= \frac{-8}{3} = -\frac{8}{3}$$

The slope of the line is $\left(-\frac{8}{3}\right)$.

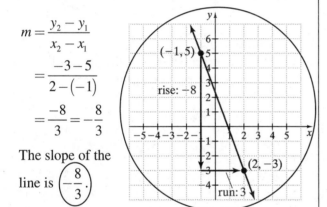

Your turn:

2. Find the slope of the line through $(-1, 5)$ and $(6, -2)$.

Review this example:

3. Find the slope and the y-intercept of the line whose equation is $3x - 4y = 4$.

Write the equation in slope-intercept form by solving for y.

$3x - 4y = 4$

$\quad -4y = -3x + 4$

$\quad \dfrac{-4y}{-4} = \dfrac{-3x}{-4} + \dfrac{4}{-4}$

$\quad\quad y = \dfrac{3}{4}x - 1$

The coefficient of x, $\left(\dfrac{3}{4}\right)$, is the slope, and the y-intercept is $\left((0, -1)\right)$.

Your turn:

4. Find the slope of the line $2x - 3y = 10$.

Section 3.4 Slope and Rate of Change

Review this example:

5. Find the slope of the line $x = 5$.

The graph of $x = 5$ is a vertical line with x-intercept $(5,0)$. To find the slope, find two ordered pair solutions of $x = 5$. Solutions must have an x-value of 5. Let's use points $(5,0)$ and $(5,4)$, which are on the line.

$$m = \frac{y_2 - y_1}{x_2 - x_1} = \frac{4-0}{5-5} = \frac{4}{0}$$

Since $\frac{4}{0}$ is undefined, the slope of the line $x = 5$ is (undefined.)

Your turn:

6. Find the slope of the line $y = -3$.

Review this example:

7. Determine whether the pair of lines is parallel, perpendicular, or neither.

$$y = -\frac{1}{5}x + 1$$
$$2x + 10y = 3$$

The slope of the line $y = -\frac{1}{5}x + 1$ is $-\frac{1}{5}$. Find the slope of the second line by solving its equation for y.

$$2x + 10y = 3$$
$$10y = -2x + 3$$
$$y = \frac{-2}{10}x + \frac{3}{10}$$
$$y = -\frac{1}{5}x + \frac{3}{10}$$

The slope of this line is $-\frac{1}{5}$ also. Since the lines have the same slope and different y-intercepts, they are (parallel.)

Your turn:

8. Determine whether the pair of lines is parallel, perpendicular, or neither.

$$y = \frac{2}{9}x + 3$$
$$y = -\frac{2}{9}x$$

Review this example:

9. The following graph shows annual food and drink sales y (in billions of dollars) for year x. Find the slope of the line and attach the proper units for the rate of change. Then write a sentence explaining the meaning of slope in this application.

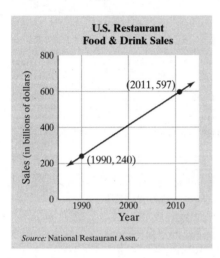

Source: National Restaurant Assn.

Use $(1990, 240)$ and $(2011, 597)$ to calculate slope.

$$m = \frac{597 - 240}{2011 - 1990} = \frac{357}{21} = \boxed{\frac{17 \text{ billion dollars}}{1 \text{ year}}}$$

This means that the rate of change of restaurant food and drink sales increases by 17 billion dollars every 1 year, or $17 billion per year.

Your turn:

10. Find the slope of the line and write the slope as a rate of change. Don't forget the proper units.

The graph shows the total cost y (in dollars) of owning and operating a compact car where x is the number of miles driven.

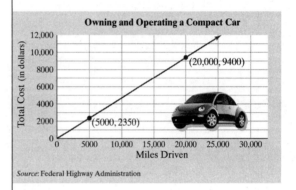

Source: Federal Highway Administration

101

Section 3.4 Slope and Rate of Change

	Answer	Text Ref	Video Ref		Answer	Text Ref	Video Ref
1	$-\dfrac{8}{3}$;	Ex 1, p. 208		6	$m=0$		Sec 3.4, 8/11
2	-1		Sec 3.4, 2/11	7	parallel	Ex 8a, p. 213	
3	$m=\dfrac{3}{4}$; $(0,-1)$	Ex 5, p. 210		8	neither		Sec 3.4, 9/11
4	$m=\dfrac{2}{3}$		Sec 3.4, 6/11	9	$m=17$; restaurant food and drink sales increase by $17 billion per year.	Ex 10, p. 215	
5	undefined	Ex 7, p. 211		10	$m=0.47$; It costs $0.47 per mile to own and operate a compact car.		Sec 3.4, 11/11

☐ **Next, insert your homework.** Make sure you attempt all exercises asked of you and show all work, as in the exercises above. Check your answers if possible. Clearly mark any exercises you were unable to correctly complete so that you may ask questions later. DO NOT ERASE YOUR INCORRECT WORK. THIS IS HOW WE UNDERSTAND AND EXPLAIN TO YOU YOUR ERRORS.

Section 3.5 Equations of Lines

Before Class:

☐   Read the objectives on page 222.

☐   Read the **Helpful Hint** boxes on pages 223 and 225.

☐   Complete the exercises:

1. Given the slope of a line and any point on the line, write the equation of the line using the

    _____ form.

2. Given two points on a line, what must be done first in order to write the equation of the line?

3. Nonvertical parallel lines have the _____ slope.

During Class:

☐   **Write your class notes.** Neatly write down **all** examples shown as well as key terms or phrases with definitions. If not applicable or if you were absent, watch the Lecture Series (DVD) for this section and do the same (write down the examples shown as well as key terms or phrases). Insert more paper as needed.

Class Notes/Examples	Your Notes

Answers: **1)** point-slope   **2)** Find the slope of the line.   **3)** same

Section 3.5 Equations of Lines

Class Notes (continued)

Your Notes

(Insert additional paper as needed.)

Section 3.5 Equations of Lines

Practice:

☐   Complete the Vocabulary, Readiness & Video Check on page 228.

☐   Next, complete any incomplete exercises below. Check and correct your work using the answers and references at the end of this section.

Review this example:

1. Use the slope-intercept form to graph the

equation $y = \dfrac{3}{5}x - 2$.

The slope is $\dfrac{3}{5}$ and the y-intercept is $(0, -2)$. Plot

the point $(0, -2)$. From this point, find another

point on the line using the slope $\dfrac{3}{5}$ and recalling

that slope is $\dfrac{\text{rise}}{\text{run}}$. Start at $(0, -2)$ and move 3 units

up and 5 units to the right, stopping at the point

$(5, 1)$. Draw the line through $(0, -2)$ and $(5, 1)$.

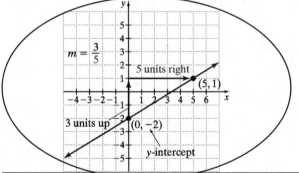

Your turn:

2. Use the slope-intercept form to graph $4x - 7y = -14$.

Review this example:

3. Find an equation of the line with y-intercept

$(0, -3)$ and slope of $\dfrac{1}{4}$.

Let $m = \dfrac{1}{4}$ and $b = -3$.

$y = mx + b$ ←   Write the equation in slope-intercept form.

$y = \dfrac{1}{4}x + (-3)$

$y = \dfrac{1}{4}x - 3$

Your turn:

4. Write an equation of the line with slope $m = -4$ and y-intercept

$\left(0, -\dfrac{1}{6}\right)$.

Section 3.5 Equations of Lines

Review this example:

5. Find an equation of the line with slope -2 that passes through $(-1,5)$. Write the equation in slope-intercept form, $y = mx + b$, and in standard form, $Ax + By = C$.

Since the slope and a point on the line are given, use point-slope form to write the equation. Let $m = -2$ and $(-1,5) = (x_1, y_1)$.

$$y - y_1 = m(x - x_1)$$
$$y - 5 = -2[x - (-1)]$$
$$y - 5 = -2(x + 1)$$
$$y - 5 = -2x - 2$$
$$y = -2x + 3 \quad \text{slope-intercept form}$$
$$2x + y = 3 \quad \quad \text{standard form}$$

Your turn:

6. Find an equation of the line with slope $m = -8$ that passes through the point $(-1, -5)$. Write the equation in the standard form, $Ax + By = C$.

Review this example:

7. Find an equation of the line through $(2,5)$ and $(-3,4)$. Write the equation in standard form.

First, use the given points to find the slope of the line.
$$m = \frac{4 - 5}{-3 - 2} = \frac{-1}{-5} = \frac{1}{5}$$

Next use the slope and either one of the given points to write the equation in point-slope form. Let $x_1 = 2$, $y_1 = 5$, and $m = \frac{1}{5}$.

$$y - y_1 = m(x - x_1)$$
$$y - 5 = \frac{1}{5}(x - 2)$$
$$5(y - 5) = 5 \cdot \frac{1}{5}(x - 2)$$
$$5y - 25 = x - 2$$
$$-x + 5y - 25 = -2$$
$$-x + 5y = 23$$

Your turn:

8. Find an equation of the line passing through $(2,3)$ and $(-1,-1)$. Write the equation in the standard form, $Ax + By = C$.

Review this example:

9. Find an equation of the line parallel to the line $y = 5$ and passing through $(-2, -3)$.

Since the graph of $y = 5$ is a horizontal line, any line parallel to it is also horizontal. The equation of a horizontal line can be written in the form $y = c$. An equation for the horizontal line passing through $(-2, -3)$ is $y = -3$.

Your turn:

10. Find an equation of the line parallel to $y = 5$ that passes through $(1, 2)$.

Review this example:

11. A web-based T-shirt company has learned that by pricing a clearance-sale T-shirt at $6, sales will reach 2000 T-shirts per day. Raising the price to $8 will cause the sales to fall to 1500 T-shirts per day.

   a. Assume that the relationship between sales price and number of T-shirts sold is linear and write an equation in slope-intercept form describing this relationship.

   b. Predict the daily sales of T-shirts if the price is $7.50.

a. Using the given information to write two ordered pairs in the form (sales price, number sold), the ordered pairs are (6, 2000) and (8, 1500). Find the slope of the line that contains these points.

$$m = \frac{2000 - 1500}{6 - 8} = \frac{500}{-2} = -250$$

Use (6, 2000) to write the equation in point-slope form.

$$y - y_1 = m(x - x_1)$$
$$y - 2000 = -250(x - 6)$$
$$y - 2000 = -250x + 1500$$
$$y = -250x + 3500$$

b. To predict the sales if the price is $7.50, find y when $x = 7.50$.

$$y = -250x + 3500$$
$$y = -250(7.50) + 3500$$
$$y = -1875 + 3500$$
$$y = 1625 \qquad \leftarrow \text{If the price is \$7.50, sales will reach 1625 T-shirts per day.}$$

Your turn:

12. A rock is dropped from the top of a 400-foot cliff. After 1 second, the rock is traveling 32 feet per second. After 3 seconds, the rock is traveling 96 feet per second.

   a. Assume that the relationship between time and speed is linear and write an equation describing this relationship. Use ordered pairs of the form (time, speed).

   b. Use the equation to determine the speed of the rock 4 seconds after it was dropped.

Section 3.5 Equations of Lines

	Answer	Text Ref	Video Ref		Answer	Text Ref	Video Ref
1		Ex 1, p. 223		7	$-x + 5y = 23$	Ex 5, p. 225	
2		Sec 3.5, 2/8	8	$-4x + 3y = 1$ or $4x - 3y = -1$		Sec 3.5, 5/8
3	$y = \dfrac{1}{4}x - 3$	Ex 3, p. 223	9	$y = -3$	Ex 7, p. 226	
4	$y = -4x - \dfrac{1}{6}$		Sec 3.5, 3/8	10	$y = 2$		Sec 3.5, 6/8
5	$y = -2x + 3,$ $2x + y = 3$	Ex 4, p. 224	11	a. $y = -250x + 3500$ b. 1625 T-shirts/day	Ex 8, p. 226	
6	$8x + y = -13$		Sec 3.5, 4/8	12	a. $s = 32t$ b. 128 ft/sec		Sec 3.5, 8/8

☐   **Next, insert your homework.** Make sure you attempt all exercises asked of you and show all work, as in the exercises above. Check your answers if possible. Clearly mark any exercises you were unable to correctly complete so that you may ask questions later. DO NOT ERASE YOUR INCORRECT WORK. THIS IS HOW WE UNDERSTAND AND EXPLAIN TO YOU YOUR ERRORS.

Before Class:

☐   Read the objectives on page 231.

☐   Read the **Helpful Hint** boxes on page 237.

☐   Complete the exercises:

  1. A _____ is a set of ordered pairs that assigns to each x-value exactly one y-value.

  2. According to the vertical line test, if a vertical line intersects a graph more than once, is the graph the graph of a function?

  3. All _____ equations are functions except those of the form $x = c$, which are vertical lines.

During Class:

☐   **Write your class notes.** Neatly write down **all** examples shown as well as key terms or phrases with definitions. If not applicable or if you were absent, watch the Lecture Series (DVD) for this section and do the same (write down the examples shown as well as key terms or phrases). Insert more paper as needed.

Class Notes/Examples	**Your Notes**

Answers: **1)** function   **2)** no   **3)** linear

Section 3.6 Functions

Class Notes (continued)	Your Notes

(Insert additional paper as needed.)

Practice:

☐   Complete the Vocabulary, Readiness & Video Check on page 239.

☐   Next, complete any incomplete exercises below. Check and correct your work using the answers and references at the end of this section.

Review this example:	**Your turn:**
1. Find the domain and the range of the relation $\{(0,2),(3,3),(-1,0),(3,-2)\}$	**2.** Find the domain and range of the relation $\{(0,-2),(1,-2),(5,-2)\}$.

The domain is the set of all *x*-values or $\{-1,0,3\}$.

The range is the set of all *y*-values or $\{-2,0,2,3\}$

Review this example:

3. Which of the following relations are also functions?

     a. $\{(-1,1),(2,3),(7,3),(8,6)\}$

Although the ordered pairs $(2,3)$ and $(7,3)$ have the same *y*-value, each *x*-value is assigned to only one *y*-value so this set of ordered pairs is a function.

     b. $\{(0,-2),(1,5),(0,3),(7,7)\}$

The *x*-value 0 is assigned to two *y*-values, -2 and 3, so this set of ordered pairs is not a function.

Your turn:

4. Determine whether each relation is also a function.

     a. $\{(1,1),(2,2),(-3,-3),(0,0)\}$

     b. $\{(-1,0),(-1,6),(-1,8)\}$

Review this example:

5. Which graph is the graph of a function?

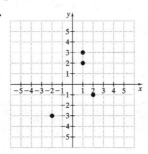

Your turn:

6. Determine whether the graph is the graph of a function.

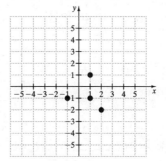

a. Each *x*-coordinate has exactly one *y*-coordinate, so this is the graph of a function.

b. The *x*-coordinate 1 is paired with two *y*-coordinates, 2 and 3, so this is not the graph of a function.

Section 3.6 Functions

Review this example:	**Your turn:**
7. Given $g(x) = x^2 - 3$, find the following.	**8.** Given $f(x) = x^2 + 2$, find
a. $g(2)$ b. $g(-2)$ c. $g(0)$	$f(-2), f(0)$, and $f(3)$.

a. $g(x) = x^2 - 3$

$g(2) = 2^2 - 3 = 4 - 3 = \boxed{1}$

b. $g(x) = x^2 - 3$

$g(-2) = (-2)^2 - 3 = 4 - 3 = \boxed{1}$

c. $g(x) = x^2 - 3$

$g(0) = 0^2 - 3 = 0 - 3 = \boxed{-3}$

Review this example:	**Your turn:**
9. Find the domain and range of the function graphed.	**10.** Find the domain and range of the relation graphed.

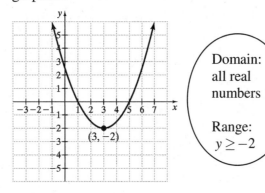

Domain: all real numbers

Range: $y \geq -2$

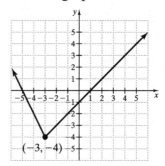

	Answer	Text Ref	Video Ref		Answer	Text Ref	Video Ref
1	$\{-1,0,3\}\,;\{-2,0,2,3\}$	Ex 1, p. 232		**6**	no		Sec 3.6, 4/13
2	$\{0,1,5\};\{-2\}$		Sec 3.6, 1/13	**7**	a. 1 b. 1 c. −3	Ex 7, p. 237	
3	a. function b. not a function	Ex 2, p. 232		**8**	6, 2, 11		Sec 3.6, 10/13
4	a. yes b. no		Sec 3.6, 2–3/13	**9**	domain: $(-\infty, \infty)$ range: $[-2, \infty)$	Ex 9b, p. 238	
5	a. function b. not a function	Ex 3, p. 233		**10**	domain: $(-\infty, \infty)$ range: $[-4, \infty)$		Sec 3.6, 11/13

☐   **Next, insert your homework.** Make sure you attempt all exercises asked of you and show all work, as in the exercises above. Check your answers if possible. Clearly mark any exercises you were unable to correctly complete so that you may ask questions later. DO NOT ERASE YOUR INCORRECT WORK. THIS IS HOW WE UNDERSTAND AND EXPLAIN TO YOU YOUR ERRORS.

Preparing for the Chapter 3 Test

Start preparing for your Chapter 3 Test as soon as possible. Pay careful attention to any instructor discussion about this test, especially discussion on what sections you will be responsible for, etc.

☐ Work the Chapter 3 Vocabulary Check on page 243.

☐ Read your Class Notes/Examples for each section covered on your Chapter 3 Test. Look for any unresolved questions you may have.

☐ Complete as many of the Chapter 3 Review exercises as possible (page 247). Remember, the odd answers are in the back of your text.

☐ **Most important:** Place yourself in "test" conditions (see below) and work the Chapter 3 Test (page 250) as a practice test the day before your actual test. To honestly assess how you are doing, try the following:
- Work on a few blank sheets of paper.
- Give yourself the same amount of time you will be given for your actual test.
- Complete this Chapter 3 Practice Test without using your notes or your text.
- If you have any time left after completing this practice test, check your work and try to find any errors on your own.
- Once done, use the back of your book to check ALL answers.
- Try to correct any errors on your own.
- Use the Chapter Test Prep Video (CTPV) to correct any errors you were unable to correct on your own. You can find these videos in the Interactive DVD Lecture Series, in MyMathLab, and on YouTube. Search Martin-Gay Beginning Algebra and click "Channels."

I wish you the best of luck….Elayn Martin-Gay

Section 4.1 Solving Systems of Linear Equations by Graphing

Before Class:

☐   Read the objectives on page 255.

☐   Read the **Helpful Hint** boxes on page 256.

☐   Complete the exercises:

1.   What does the graph of an inconsistent system of linear equations look like?

2.   What does the graph of a system of dependent linear equations look like?

3.   What does the graph of a consistent system of independent linear equations look like?

During Class:

☐   **Write your class notes.** Neatly write down **all** examples shown as well as key terms or phrases with definitions. If not applicable or if you were absent, watch the Lecture Series (DVD) for this section and do the same (write down the examples shown as well as key terms or phrases). Insert more paper as needed.

Class Notes/Examples	**Your Notes**

Answers: **1)** parallel lines   **2)** a single line   **3)** lines that intersect in exactly one point

Section 4.1 Solving Systems of Linear Equations by Graphing

Class Notes (continued)	**Your Notes**

(Insert additional paper as needed.)

Section 4.1 Solving Systems of Linear Equations by Graphing

Practice:

☐   Complete the Vocabulary, Readiness & Video Check on page 260.

☐   Next, complete any incomplete exercises below. Check and correct your work using the answers and references at the end of this section.

Review this example:

1. Determine whether $(-1, 2)$ is a solution of the system

$$\begin{cases} x + 2y = 3 \\ 4x - y = 6 \end{cases}$$

We replace x with -1 and y with 2 in both equations.

$$x + 2y = 3 \qquad\qquad 4x - y = 6$$
$$-1 + 2(2) \overset{?}{=} 3 \qquad 4(-1) - 2 \overset{?}{=} 6$$
$$-1 + 4 \overset{?}{=} 3 \qquad\qquad -4 - 2 \overset{?}{=} 6$$
$$3 = 3 \text{ True} \qquad -6 = 6 \text{ False}$$

$(-1, 2)$ is not a solution of the second equation, so it is not a solution of the system.

Your turn:

2. Determine whether each ordered pair is a solution of the system of linear equations.

$$\begin{cases} 3x - y = 5 \\ x + 2y = 11 \end{cases}$$

a. $(3, 4)$

b. $(0, -5)$

Review this example:

3. Solve the system of equations by graphing.

$$\begin{cases} -x + 3y = 10 \\ x + y = 2 \end{cases}$$

On a single set of axes, graph each linear equation.

$-x + 3y = 10$		$x + y = 2$	
x	y	x	y
0	10/3	0	2
-4	2	2	0
2	4	1	1

(solution continued on the next page)

Your turn:

4. Solve the system of equations by graphing.

$$\begin{cases} 2x + y = 0 \\ 3x + y = 1 \end{cases}$$

Section 4.1 Solving Systems of Linear Equations by Graphing

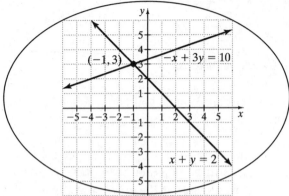

The two lines appear to intersect at the point $(-1,3)$. To check, we replace x with -1 and y with 3 in both equations.

$$-x+3y=10$$

$$-(-1)+3(3)\overset{?}{=}10 \qquad\qquad x+y=2$$

$$1+9\overset{?}{=}10 \qquad\qquad -1+3\overset{?}{=}2$$

$$10=10 \ \ \text{True} \qquad\qquad 2=2 \ \ \text{True}$$

$(-1,3)$ checks, so it is the solution of the system.

Review this example:	**Your turn:**

5. Solve the system of equations by graphing.

$$\begin{cases} 2x+y=7 \\ 2y=-4x \end{cases}$$

Graph the two lines in the system.

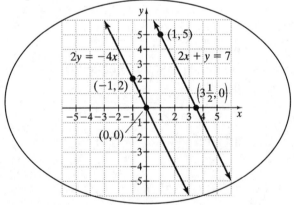

The lines *appear* to be parallel. To confirm this, write both equations in slope-intercept form by solving each equation for y. Both equations have the same slope, -2, but different y-intercepts.

There is no solution of the system.

6. Solve the system of equations by graphing.

$$\begin{cases} x+y=5 \\ x+y=6 \end{cases}$$

Copyright © 2013 Pearson Education, Inc.

Section 4.1 Solving Systems of Linear Equations by Graphing

Review this example:

7. Solve the system of equations by graphing.

$$\begin{cases} x - y = 3 \\ -x + y = -3 \end{cases}$$

Graph each line.

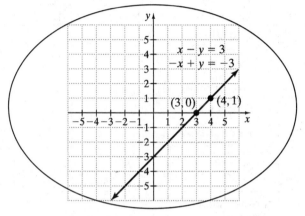

The graphs *appear* to be identical. To confirm this, write each equation in slope-intercept form. The equations are identical and so must be their graphs.

Thus, there is an infinite number of solutions of the system.

Your turn:

8. Solve the system of equations by graphing.

$$\begin{cases} 6x - y = 4 \\ \dfrac{1}{2}y = -2 + 3x \end{cases}$$

Review this example:

9. Without graphing, determine the number of solutions of the system.

$$\begin{cases} 3x - y = 4 \\ x + 2y = 8 \end{cases}$$

Write each equation in slope-intercept form.

$3x - y = 4$ $x + 2y = 8$

$\quad 3x = y + 4$ $x = -2y + 8$

$3x - 4 = y$ $x - 8 = -2y$

$$\frac{x}{-2} - \frac{8}{-2} = \frac{-2y}{-2}$$

$$-\frac{1}{2}x + 4 = y$$

The slope of the second line is $-\dfrac{1}{2}$; the slope of the first line is 3. The slopes are not equal, so the lines are neither parallel nor identical and must intersect. This system has one solution.

Your turn:

10. $\begin{cases} 4x + y = 24 \\ x + 2y = 2 \end{cases}$

   a. Are the graphs of the equations identical lines, parallel lines, or lines intersecting at a single point?

   b. How many solutions does the system have?

Section 4.1 Solving Systems of Linear Equations by Graphing

	Answer	Text Ref	Video Ref		Answer	Text Ref	Video Ref
1	not a solution	Ex 2, p. 255		6	no solution		Sec 4.1, 3/7
2	a. yes b. no		Sec 4.1, 1/7	7	infinite number of solutions	Ex 6, p. 258	
3		Ex 3, p. 256		8	infinite number of solutions		Sec 4.1, 4/7
4			Sec 4.1, 2/7	9	one solution	Ex 8, pp. 259–260	
5	no solution	Ex 5, pp. 257–258		10	intersecting, one solution		Sec 4.1, 5/7

☐ **Next, insert your homework.** Make sure you attempt all exercises asked of you and show all work, as in the exercises above. Check your answers if possible. Clearly mark any exercises you were unable to correctly complete so that you may ask questions later. DO NOT ERASE YOUR INCORRECT WORK. THIS IS HOW WE UNDERSTAND AND EXPLAIN TO YOU YOUR ERRORS.

Section 4.2 Solving Systems of Linear Equations by Substitution

Before Class:

☐   Read the objective on page 263.

☐   Read the **Helpful Hint** boxes on pages 263, 265, and 266.

☐   Complete the exercises:

    1.   Read the Solving a System of Two Linear Equations by the Substitution Method box on page 265. What is the first step?

    2.   What is the last step in solving a system of two linear equations by the substitution method?

During Class:

☐   **Write your class notes.** Neatly write down **all** examples shown as well as key terms or phrases with definitions. If not applicable or if you were absent, watch the Lecture Series (DVD) for this section and do the same (write down the examples shown as well as key terms or phrases). Insert more paper as needed.

Class Notes/Examples	Your Notes

Answers: **1)** Solve one of the equations for one of its variables.   **2)** Check the proposed solution in the original system.

Section 4.2 Solving Systems of Linear Equations by Substitution

Class Notes (continued)

(Insert additional paper as needed.)

Section 4.2 Solving Systems of Linear Equations by Substitution

Practice:

☐ Complete the Vocabulary, Readiness & Video Check on page 268.

☐ Next, complete any incomplete exercises below. Check and correct your work using the answers and references at the end of this section.

Review this example:

1. Solve the system: $\begin{cases} 5x - y = -2 \\ y = 3x \end{cases}$

The second equation is solved for y in terms of x. We substitute $3x$ for y in the first equation and solve for x.

$$5x - y = -2$$
$$5x - (3x) = -2$$
$$2x = -2$$
$$x = -1$$

The x-value of the ordered pair solutions is -1. To find the corresponding y-value, replace x with -1 in the second equation.

$$y = 3x$$
$$y = 3(-1) = -3$$

Check that the solution of the system is $(-1, -3)$.

Your turn:

2. Solve the system by the substitution method.

$$\begin{cases} x + y = 6 \\ y = -3x \end{cases}$$

Review this example:

3. Solve the system:

$$\begin{cases} x + 2y = 7 \\ 2x + 2y = 13 \end{cases}$$

Solve the first equation for x: $\quad x + 2y = 7$
$$x = 7 - 2y$$

Substitute $7 - 2y$ for x in the second equation and solve for y.

$$2x + 2y = 13$$
$$2(7 - 2y) + 2y = 13$$
$$14 - 4y + 2y = 13$$
$$14 - 2y = 13$$
$$-2y = -1$$
$$y = \frac{1}{2}$$

(solution continued on the next page)

Your turn:

4. Solve the system by the substitution method.

$$\begin{cases} 3x - y = 1 \\ 2x - 3y = 10 \end{cases}$$

Section 4.2 Solving Systems of Linear Equations by Substitution

To find x, let $y = \dfrac{1}{2}$ in the equation $x = 7 - 2y$.

$x = 7 - 2y$

$x = 7 - 2\left(\dfrac{1}{2}\right)$

$x = 7 - 1 = 6$

The solution is $\left(6, \dfrac{1}{2}\right)$. Check the solution in both

equations of the original system.

Review this example:

5. Use substitution to solve the system.

$$\begin{cases} 6x + 12y = 5 \\ -4x - 8y = 0 \end{cases}$$

Solve the second equation for y: $-4x - 8y = 0$

$$-8y = 4x$$

$$y = -\dfrac{1}{2}x$$

Replace y with $-\dfrac{1}{2}x$ in the first equation.

$6x + 12y = 5$

$6x + 12\left(-\dfrac{1}{2}x\right) = 5$

$6x + (-6x) = 5$

$0 = 5$ False

The system has no solution.

Your turn:

6. Solve the system by the substitution method.

$$\begin{cases} 3x + 6y = 9 \\ 4x + 8y = 16 \end{cases}$$

	Answer	Text Ref	Video Ref		Answer	Text Ref	Video Ref
1	$(-1,-3)$	Ex 2, p. 264		4	$(-1,-4)$		Sec 4.2, 2/4
2	$(-3,9)$		Sec 4.2, 1/4	5	no solution	Ex 6, pp. 267–268	
3	$\left(6, \dfrac{1}{2}\right)$	Ex 3, p. 265		6	no solution		Sec 4.2, 3/4

☐   **Next, insert your homework.** Make sure you attempt all exercises asked of you and show all work, as in the exercises above. Check your answers if possible. Clearly mark any exercises you were unable to correctly complete so that you may ask questions later. DO NOT ERASE YOUR INCORRECT WORK. THIS IS HOW WE UNDERSTAND AND EXPLAIN TO YOU YOUR ERRORS.

Section 4.3 Solving Systems of Linear Equations by Addition

Before Class:

☐   Read the objective on page 270.

☐   Read the **Helpful Hint** boxes on pages 270 and 271, and 272.

☐   Complete the exercises:

1. The addition method is based on the addition property of_____:
    adding equal quantities to both sides of an equation does not change the solution of the
    equation.

2. What is the goal when solving a system of equations by the addition method?

3. Read the Solving a System of Two Linear Equations by the Addition Method box on
    page 273. What is the last step?

During Class:

☐   **Write your class notes.** Neatly write down **all** examples shown as well as key terms or
     phrases with definitions. If not applicable or if you were absent, watch the Lecture Series
     (DVD) for this section and do the same (write down the examples shown as well as key terms
     or phrases). Insert more paper as needed.

Class Notes/Examples	**Your Notes**

Answers: **1)** equality   **2)** to eliminate a variable when adding the equations   **3)** Check the
proposed solution in the original system.

Section 4.3 Solving Systems of Linear Equations by Addition

Class Notes (continued)

Your Notes

(Insert additional paper as needed.)

Section 4.3 Solving Systems of Linear Equations by Addition

Practice:

☐   Complete the Vocabulary, Readiness & Video Check on page 274–275.

☐   Next, complete any incomplete exercises below. Check and correct your work using the answers and references at the end of this section.

Review this example:

1. Solve the system: $\begin{cases} x+y=7 \\ x-y=5 \end{cases}$

Add the left sides of the equations together and the right sides of the equations together. This eliminates the variable y and gives us an equation in one variable, x. Then solve for x.

$x+y=7$

$\underline{x-y=5}$

$2x \quad =12$

$\quad x=6$

To find the corresponding y-value, let $x=6$ in either equation of the system.

$x+y=7$

$6+y=7$

$\quad y=1$

The solution is $(6,1)$. Check this in both equations.

Your turn:

2. Solve by the addition method.

$\begin{cases} x-2y=8 \\ -x+5y=-17 \end{cases}$

Review this example:

3. Solve the system: $\begin{cases} 3x-2y=2 \\ -9x+6y=-6 \end{cases}$

First multiply both sides of the first equation by 3. Then add the resulting equations.

$\begin{cases} 3(3x-2y)=3(2) \\ -9x+6y=-6 \end{cases}$ simplifies to $\begin{cases} 9x-6y=6 \\ \underline{-9x+6y=-6} \\ \quad 0=0 \end{cases}$

Both variables are eliminated and we have 0 = 0, a true statement.

The system has an infinite number of solutions.

Your turn:

4. Solve by the addition method.

$\begin{cases} 3x-2y=7 \\ 5x+4y=8 \end{cases}$

Section 4.3 Solving Systems of Linear Equations by Addition

Review this example:

5. Solve the system: $\begin{cases} -x - \dfrac{y}{2} = \dfrac{5}{2} \\ \dfrac{x}{6} - \dfrac{y}{2} = 0 \end{cases}$

Clear each equation of fractions. Multiply both sides of the first equation by the LCD 2 and both sides of the second equation by the LCD 6.

$\begin{cases} 2\left(-x - \dfrac{y}{2}\right) = 2\left(\dfrac{5}{2}\right) \\ 6\left(\dfrac{x}{6} - \dfrac{y}{2}\right) = 6(0) \end{cases}$ simplifies to $\begin{cases} -2x - y = 5 \\ x - 3y = 0 \end{cases}$

Multiply both sides of the second equation by 2 to eliminate the variable x.

$\begin{cases} -2x - y = 5 \\ 2(x - 3y) = 2(0) \end{cases}$ simplifies to $\begin{cases} -2x - y = 5 \\ \underline{2x - 6y = 0} \\ -7y = 5 \end{cases}$

Add the equations in $\rightarrow$
the simplified system. $y = -\dfrac{5}{7}$

To find x, go back to the simplified system and multiply the first equation in the simplified system by -3 to eliminate the variable y and solve for x.

$\begin{cases} -3(-2x - y) = -3(5) \\ x - 3y = 0 \end{cases}$ simplifies to $\begin{cases} 6x + 3y = -15 \\ \underline{x - 3y = 0} \\ 7x = -15 \end{cases}$

$x = -\dfrac{15}{7}$

The solution is $\left(-\dfrac{15}{7}, -\dfrac{5}{7}\right)$.

Your turn:

6. Solve by the addition method. You may want to first clear each equation of fractions.

$\begin{cases} \dfrac{x}{3} - y = 2 \\ -\dfrac{x}{2} + \dfrac{3y}{2} = -3 \end{cases}$

	Answer	Text Ref	Video Ref		Answer	Text Ref	Video Ref
1	$(6,1)$	Ex 1, p. 270		4	$\left(2, -\dfrac{1}{2}\right)$		Sec 4.3, 2/4
2	$(2,-3)$		Sec 4.3, 1/4	5	$\left(-\dfrac{15}{7}, -\dfrac{5}{7}\right)$	Ex 6, pp. 273–274	
3	infinite number of solutions	Ex 4, p. 272		6	infinite number of solutions		Sec 4.3, 3/4

☐   **Next, insert your homework.** Make sure you attempt all exercises asked of you and show all work, as in the exercises above. Check your answers if possible. Clearly mark any exercises you were unable to correctly complete so that you may ask questions later. DO NOT ERASE YOUR INCORRECT WORK. THIS IS HOW WE UNDERSTAND AND EXPLAIN TO YOU YOUR ERRORS.

Section 4.4 Systems of Linear Equations and Problem Solving

Before Class:

☐   Read the objective on page 278.

☐   Write the four steps to solving word problems using a system of two equations.

1.

2.

3.

4.

During Class:

☐   **Write your class notes.** Neatly write down **all** examples shown as well as key terms or phrases with definitions. If not applicable or if you were absent, watch the Lecture Series (DVD) for this section and do the same (write down the examples shown as well as key terms or phrases). Insert more paper as needed.

Class Notes/Examples	Your Notes

Answers: **1)** Understand the problem.   **2)** Translate the problem into two equations.   **3)** Solve the system of equations.   **4)** Interpret the results.

Section 4.4 Systems of Linear Equations and Problem Solving

Class Notes (continued)	**Your Notes**

(Insert additional paper as needed.)

Section 4.4 Systems of Linear Equations and Problem Solving

Practice:

☐   Complete the Vocabulary, Readiness & Video Check on page 284.

☐   Next, complete any incomplete exercises below. Check and correct your work using the answers and references at the end of this section.

Review this example:

1. Find two numbers whose sum is 37 and whose difference is 21.

UNDERSTAND. Read and reread the problem. Let x = first number, and y = second number.

TRANSLATE.
$x + y = 37$
$x - y = 21$

SOLVE. Solve the system $\begin{cases} x + y = 37 \\ x - y = 21 \end{cases}$.

The coefficients of the variable y are opposites. Solve by the addition method.

$$\begin{array}{r} x + y = 37 \\ \underline{x - y = 21} \\ 2x \quad\;\; = 58 \end{array}$$

$$x = \frac{58}{2} = 29$$

Let $x = 29$ in the first equation to find y.
$x + y = 37$
$29 + y = 37$
$y = 8$

INTERPRET. The solution of the system is $(29, 8)$.

Check: Notice that the sum of 29 and 8 is 37 and their difference is 21.

The numbers are 29 and 8.

Your turn:

2. Two numbers total 83 and have a difference of 17. Find the two numbers.

Section 4.4 Systems of Linear Equations and Problem Solving

Review this example:

3. Eric needs 10 liters of a 20% saline solution (salt water). The only mixtures on hand are a 5% saline solution and a 25% saline solution. How much of each solution should he mix to produce the 20% solution?

UNDERSTAND. Read and reread the problem. Let x = number of liters of 5% solution, and y = number of liters of 25% solution.

	Concentration Rate	Liters of Solution	Liters of Pure Salt
First Solution	5%	x	$0.05x$
Second Solution	25%	y	$0.25y$
Mixture Needed	20%	10	$(0.20)(10)$

TRANSLATE.

$$\begin{aligned} x + y &= 10 \\ 0.05x + 0.25y &= (0.20)(10) \end{aligned} \rightarrow \begin{cases} x + y = 10 \\ 0.05x + 0.25y = 2 \end{cases}$$

SOLVE. Solve the system by the addition method.

$$\begin{cases} -25x - 25y = -250 & \leftarrow \text{Multiply by } -25. \\ \underline{5x + 25y = 200} & \leftarrow \text{Multiply by 100.} \end{cases}$$

$$-20x = -50 \quad \text{or} \quad x = 2.5$$

Let $x = 2.5$ in the first equation of the original system and solve for y: $\quad x + y = 10$

$$2.5 + y = 10$$

$$y = 7.5$$

INTERPRET.

Eric needs to mix 2.5 liters of 5% saline solution with 7.5 liters of 25% saline solution.

Your turn:

4. Doreen Schmidt is a chemist with Gemco Pharmaceutical. She needs to prepare 12 liters of a 9% hydrochloric acid solution. Find the amount of a 4% solution and the amount of a 12% solution she should mix to get this solution.

Concentration Rate	Liters of Solution	Liters of Pure Acid
0.04	x	$0.04x$
0.12	y	
0.09	12	

	Answer	Text Ref	Video Ref		Answer	Text Ref	Video Ref
1	29 and 8	Ex 1, pp. 278–279		**3**	5% solution: 2.5 L, 25% solution: 7.5 L	Ex 4, pp. 282–283	
2	33 and 50		Sec 4.4, 1/5	**4**	12% solution: 7.5 L, 4% solution: 4.5 L		Sec 4.4, 4/5

☐ **Next, insert your homework.** Make sure you attempt all exercises asked of you and show all work, as in the exercises above. Check your answers if possible. Clearly mark any exercises you were unable to correctly complete so that you may ask questions later. DO NOT ERASE YOUR INCORRECT WORK. THIS IS HOW WE UNDERSTAND AND EXPLAIN TO YOU YOUR ERRORS.

Section 4.5 Graphing Linear Inequalities

Before Class:

☐   Read the objectives on page 288.

☐   Read the **Helpful Hint** box on page 291

☐   Complete the exercises:

1.   The definition of a linear inequality is the same as the definition of a linear equation

except that the equal sign is replaced with _____ .

4.   An ordered pair is a(n) _____ of an inequality in *x* and *y* if
replacing the variables by the coordinates of the ordered pair results in a true statement.

5.   When graphing an inequality, make sure the test point is substituted into the

_____ .

During Class:

☐   **Write your class notes.** Neatly write down **all** examples shown as well as key terms or
phrases with definitions. If not applicable or if you were absent, watch the Lecture Series
(DVD) for this section and do the same (write down the examples shown as well as key terms
or phrases). Insert more paper as needed.

Class Notes/Examples	**Your Notes**

Answers: **1)** an inequality sign   **2)** solution   **3)** original inequality

Section 4.5 Graphing Linear Inequalities

Class Notes (continued)

Your Notes

(Insert additional paper as needed.)

Section 4.5 Graphing Linear Inequalities

Practice:

☐   Complete the Vocabulary, Readiness & Video Check on page 293.

☐   Next, complete any incomplete exercises below. Check and correct your work using the answers and references at the end of this section.

Review this example:	**Your turn:**
1. Graph: $2x - y \geq 3$	**2.** Graph the inequality: $y \geq 2x$

Step 1. We graph the boundary line by graphing $2x - y = 3$. We draw this line as a solid line because the inequality sign is $\geq$, and thus the points on the line are solutions of $2x - y \geq 3$.

Step 2. $(0,0)$ is a convenient test point since it is not on the boundary line. We substitute 0 for x and 0 for y into the original inequality.

$$2x - y \geq 3$$
$$2(0) - 0 \geq 3$$
$$0 \geq 3 \quad \text{False}$$

Step 3. Since the statement is false, no point in the half-plane containing $(0,0)$ is a solution. Therefore, we shade the half-plane that does not contain $(0,0)$. Every point in the shaded half-plane and every point on the boundary line is a solution of $2x - y \geq 3$.

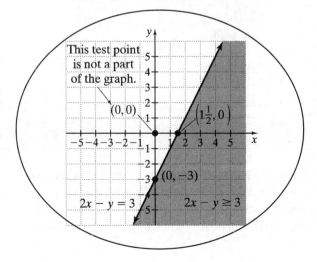

Section 4.5 Graphing Linear Inequalities

Review this example:

3. Graph: $x > 2y$

Step 1. Find the boundary line by graphing $x = 2y$.
        The boundary line is dashed since the
        inequality symbol is >.

Step 2. Use $(0, 2)$ as a test point.

$$x > 2y$$

$$0 > 2(2)$$

$$0 > 4 \qquad \text{False}$$

Step 3: Shade the half-plane that does not contain
        $(0, 2)$.

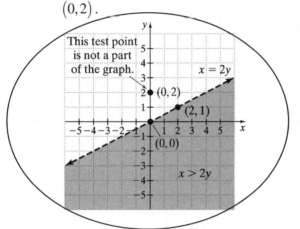

Your turn:

4. Graph the inequality: $2x + 7y > 5$

	Answer	Text Ref	Video Ref		Answer	Text Ref	Video Ref
1		Ex 3. pp. 290–291		**3**		Ex 4, p. 291	
2		Sec 4.5, 2/3		**4**		Sec 4.5, 3/3	

☐   **Next, insert your homework.** Make sure you attempt all exercises asked of you and show all work, as in the exercises above. Check your answers if possible. Clearly mark any exercises you were unable to correctly complete so that you may ask questions later. DO NOT ERASE YOUR INCORRECT WORK. THIS IS HOW WE UNDERSTAND AND EXPLAIN TO YOU YOUR ERRORS.

Section 4.6 Systems of Linear Inequalities

Before Class:

☐ Read the objective on page 295.

☐ Complete the exercises:

1. What is a corner point?

2. How do we know if a particular point on a graph is a solution of a system of linear inequalities?

During Class:

☐ **Write your class notes.** Neatly write down **all** examples shown as well as key terms or phrases with definitions. If not applicable or if you were absent, watch the Lecture Series (DVD) for this section and do the same (write down the examples shown as well as key terms or phrases). Insert more paper as needed.

Class Notes/Examples	**Your Notes**

Answers: **1)** the point at which two boundary lines intersect **2)** The ordered pair must satisfy each inequality in the system.

Section 4.6 Systems of Linear Inequalities

Class Notes (continued)	**Your Notes**

(Insert additional paper as needed.)

Section 4.6 Systems of Linear Inequalities

Practice:

☐   Complete the Vocabulary, Readiness & Video Check on page 297.

☐   Next, complete any incomplete exercises below. Check and correct your work using the
       answers and references at the end of this section.

Review this example:

1. Graph the solution of the system:

$$\begin{cases} 3x \ge y \\ x + 2y \le 8 \end{cases}$$

Graph each inequality on the same set of axes. The
graph of the solution of the system is the region
contained in the graphs of both inequalities. It is
their intersection.

First graph $3x \ge y$. The boundary line is the graph
of $3x = y$. Sketch a solid boundary line since the
inequality $3x \ge y$ means $3x > y$ or $3x = y$. The
test point $(1,0)$ satisfies the inequality, so shade the
half-plane that includes $(1,0)$.

Next sketch a solid boundary line $x + 2y = 8$ on
the same set of axes. The test point $(0,0)$ satisfies
the inequality $x + 2y \le 8$, so shade the half-plane
that includes $(0,0)$.

An ordered pair solution of the system must satisfy
both inequalities. These solutions are points that lie
in both shaded regions. The solution of the system
includes parts of both boundary lines.

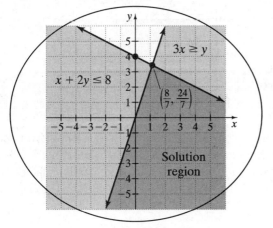

Your turn:

2. Graph the solution of the system:

$$\begin{cases} x \ge 3y \\ x + 3y \le 6 \end{cases}$$

Section 4.6 Systems of Linear Inequalities

Review this example:

3. Graph the solution of the system:

$$\begin{cases} x - y < 2 \\ x + 2y > -1 \end{cases}$$

Graph both inequalities on the same set of axes. Both boundary lines are dashed lines since the inequality symbols are < and >. The solution of the system is the region shown. The boundary lines are not part of the solution.

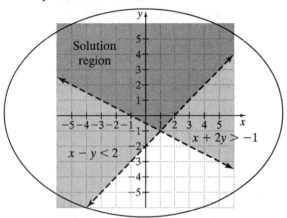

Your turn:

4. Graph the solution of the system:

$$\begin{cases} y \geq 1 \\ x < -3 \end{cases}$$

	Answer	Text Ref	Video Ref		Answer	Text Ref	Video Ref
1	$3x \geq y$, $x + 2y \leq 8$, $\left(\frac{8}{7}, \frac{24}{7}\right)$, Solution region	Ex 1, pp. 295–296		3	Solution region, $x + 2y > -1$, $x - y < 2$	Ex 2, p. 297	
2	$x \geq 3y$, $x + 3y \leq 6$	Sec 4.6, 1/2		4	$y \geq 1$, $x < -3$		Sec 4.6, 2/2

☐ **Next, insert your homework.** Make sure you attempt all exercises asked of you and show all work, as in the exercises above. Check your answers if possible. Clearly mark any exercises you were unable to correctly complete so that you may ask questions later. DO NOT ERASE YOUR INCORRECT WORK. THIS IS HOW WE UNDERSTAND AND EXPLAIN TO YOU YOUR ERRORS.

Preparing for the Chapter 4 Test

Start preparing for your Chapter 4 Test as soon as possible. Pay careful attention to any instructor discussion about this test, especially discussion on what sections you will be responsible for, etc.

☐ Work the Chapter 4 Vocabulary Check on page 299.

☐ Read your Class Notes/Examples for each section covered on your Chapter 4 Test. Look for any unresolved questions you may have.

☐ Complete as many of the Chapter 4 Review exercises as possible (page 302). Remember, the odd answers are in the back of your text.

☐ **Most important:** Place yourself in "test" conditions (see below) and work the Chapter 4 Test (page 303) as a practice test the day before your actual test. To honestly assess how you are doing, try the following:

- Work on a few blank sheets of paper.
- Give yourself the same amount of time you will be given for your actual test.
- Complete this Chapter 4 Practice Test without using your notes or your text.
- If you have any time left after completing this practice test, check your work and try to find any errors on your own.
- Once done, use the back of your book to check ALL answers.
- Try to correct any errors on your own.
- Use the Chapter Test Prep Video (CTPV) to correct any errors you were unable to correct on your own. You can find these videos in the Interactive DVD Lecture Series, in MyMathLab, and on YouTube. Search Martin-Gay Beginning Algebra and click "Channels."

I wish you the best of luck....Elayn Martin-Gay

Section 5.1 Exponents

Before Class:

☐   Read the objectives on page 307.

☐   Read the **Helpful Hint** boxes on pages 307, 308, 309, and 310.

☐   Complete the exercises:

1.   To multiply two exponential expressions with a common base, keep the

_____ and add the _____ .

2.   To raise a power to a power, keep the _____ and multiply the

_____ .

3.   To raise a product to a power, raise _____ to the power.

During Class:

☐   **Write your class notes.** Neatly write down **all** examples shown as well as key terms or phrases with definitions. If not applicable or if you were absent, watch the Lecture Series (DVD) for this section and do the same (write down the examples shown as well as key terms or phrases). Insert more paper as needed.

Class Notes/Examples	**Your Notes**

Answers: **1)** base, exponents   **2)** base, exponents   **3)** each factor

Section 5.1 Exponents

Class Notes (continued)	**Your Notes**

(Insert additional paper as needed.)

Practice:

☐ Complete the Vocabulary, Readiness & Video Check on page 315.

☐ Next, complete any incomplete exercises below. Check and correct your work using the answers and references at the end of this section.

Review this example:	Your turn:
1. Evaluate each expression for the given value of x.	**2.** Evaluate $\dfrac{2z^4}{5}$ when $z = -2$.

a. $2x^3$; x is 5   b. $\dfrac{9}{x^2}$; x is -3

a. If x is 5,   b. If x is -3,

$2x^3 = 2 \cdot (5)^3$ $\dfrac{9}{x^2} = \dfrac{9}{(-3)^2}$

$\quad = 2 \cdot (5 \cdot 5 \cdot 5)$ $= \dfrac{9}{(-3)(-3)}$

$\quad = 2 \cdot 125$

$\quad = \boxed{250}$ $= \dfrac{9}{9} = \boxed{1}$

Review this example:

3. Simplify:

a. $(2x^2)(-3x^5)$ b. $(x^2y)(x^3y^2)$

a. $(2x^2)(-3x^5) = 2 \cdot x^2 \cdot -3 \cdot x^5$

$\qquad = 2 \cdot -3 \cdot x^2 \cdot x^5$

$\qquad = \boxed{-6x^7}$

b. $(x^2y)(x^3y^2) = (x^2 \cdot x^3)(y^1 \cdot y^2)$

$\qquad = x^5 \cdot y^3 = \boxed{x^5y^3}$

Your turn:

4. Use the product rule to simplify each expression. Write the results using exponents.

a. $(5y^4)(3y)$

b. $(x^9y)(x^{10}y^5)$

Review this example:

5. Simplify each expression.

a. $(st)^4$ b. $\left(\dfrac{m}{n}\right)^7$

a. $(st)^4 = s^4 \cdot t^4 = \boxed{s^4t^4}$

b. $\left(\dfrac{m}{n}\right)^7 = \boxed{\dfrac{m^7}{n^7}}$, $n \neq 0$

Your turn:

6. Use the power of a product or quotient rule to simplify each expression.

a. $(pq)^8$

b. $\left(\dfrac{r}{s}\right)^9$

Section 5.1 Exponents

Review this example:	Your turn:
7. Simplify the quotient $\dfrac{(-3)^5}{(-3)^2}$.	**8.** Use the quotient rule to simplify $\dfrac{(-4)^6}{(-4)^3}$.

$$\dfrac{(-3)^5}{(-3)^2} = (-3)^3 = \boxed{-27}$$

Review this example:	Your turn:
9. Simplify each expression.	**10.** Simplify each expression.
a. $(-5)^0$ b. -5^0	a. $(2x)^0$
	b. $-7x^0$

a. $(-5)^0 = \boxed{1}$

b. $-5^0 = -1 \cdot 5^0 = -1 \cdot 1 = \boxed{-1}$

	Answer	Text Ref	Video Ref		Answer	Text Ref	Video Ref
1	a. 250 b. 1	Ex 2, p. 308		**6**	a. $p^8 q^8$ b. $\dfrac{r^9}{s^9}$		Sec 5.1, 17–18/30
2	$\dfrac{32}{5}$		Sec 5.1, 8/30	**7**	-27	Ex 9c, p. 313	
3	a. $-6x^7$ b. $x^5 y^3$	Ex 4, 5a, p. 309		**8**	-64		Sec 5.1, 22/30
4	a. $15y^5$ b. $x^{19} y^6$		Sec 5.1, 12–13/30	**9**	a. 1 b. -1	Ex 10c, d, p. 314	
5	a. $s^4 t^4$ b. $\dfrac{m^7}{n^7}$	Ex 7a, 8a, pp. 311–312		**10**	a. 1 b. -7		Sec 5.1, 25–26/30

☐   **Next, insert your homework.** Make sure you attempt all exercises asked of you and show all work, as in the exercises above. Check your answers if possible. Clearly mark any exercises you were unable to correctly complete so that you may ask questions later. DO NOT ERASE YOUR INCORRECT WORK. THIS IS HOW WE UNDERSTAND AND EXPLAIN TO YOU YOUR ERRORS.

Section 5.2 Adding and Subtracting Polynomials

Before Class:

☐   Read the objectives on page 318.

☐   Read the **Helpful Hint** boxes on pages 322 and 324.

☐   Complete the exercises:

   1.   What is a term?

   2.   What is the degree of a term?

   3.   What is the degree of a constant?

   4.   What are like terms?

During Class:

☐   **Write your class notes.** Neatly write down **all** examples shown as well as key terms or phrases with definitions. If not applicable or if you were absent, watch the Lecture Series (DVD) for this section and do the same (write down the examples shown as well as key terms or phrases). Insert more paper as needed.

Class Notes/Examples	Your Notes

Answers: **1)** a number or the product of a number and variables raised to powers **2)** the sum of the exponents on the variables contained in the term **3)** 0 **4)** terms that contain exactly the same variables raised to exactly the same powers

Section 5.2 Adding and Subtracting Polynomials

Class Notes (continued)	**Your Notes**

(Insert additional paper as needed.)

Section 5.2 Adding and Subtracting Polynomials

Practice:

☐ Complete the Vocabulary, Readiness & Video Check on page 325.

☐ Next, complete any incomplete exercises below. Check and correct your work using the answers and references at the end of this section.

Review this example:	**Your turn:**
1. Find the degree of the polynomial and tell whether it is a monomial, binomial, trinomial, or none of these. $$-2t^2 + 3t + 6$$ The degree of the trinomial $-2t^2 + 3t + 6$ is 2, the greatest of any of its terms.	2. Find the degree of the polynomial and determine whether it is a monomial, binomial, trinomial, or none of these. $$12x^4 y - x^2 y^2 - 12x^2 y^4$$
Review this example:	**Your turn:**
3. Find the value of the polynomial $3x^2 - 2x + 1$ when $x = -2$. Replace x with -2 and simplify. $$3x^2 - 2x + 1 = 3(-2)^2 - 2(-2) + 1$$ $$= 3(4) + 4 + 1$$ $$= 12 + 4 + 1$$ $$= 17$$	4. Find the value of the polynomial when (a) $x = 0$ and (b) $x = -1$. $$x^2 - 5x - 2$$
Review this example:	**Your turn:**
5. Simplify the polynomial by combining any like terms. $$11x^2 + 5 + 2x^2 - 7$$ $$11x^2 + 5 + 2x^2 - 7 = 11x^2 + 2x^2 + 5 - 7$$ $$= 13x^2 - 2$$	6. Simplify by combining like terms. $$0.1y^2 - 1.2y^2 + 6.7 - 1.9$$

Section 5.2 Adding and Subtracting Polynomials

Review this example:

7. Add $\left(-2x^2+5x-1\right)$ and $\left(-2x^2+x+3\right)$.

$\left(-2x^2+5x-1\right)+\left(-2x^2+x+3\right)$

$=-2x^2+5x-1-2x^2+x+3$

$=\left(-2x^2-2x^2\right)+\left(5x+1x\right)+\left(-1+3\right)$

$=\boxed{-4x^2+6x+2}$

Your turn:

8. Add: $\left(-7x+5\right)+\left(-3x^2+7x+5\right)$

Review this example:

9. Subtract $\left(5y^2+2y-6\right)$ from

$\left(-3y^2-2y+11\right)$ using the vertical format.

Arrange the polynomials in vertical format, lining up like terms.

$$\begin{array}{l} -3y^2-2y+11 \\ \underline{-\left(5y^2+2y-6\right)} \end{array} \qquad \begin{array}{l} -3y^2-2y+11 \\ \underline{-5y^2-2y+\ 6} \\ \boxed{-8y^2-4y+17} \end{array}$$

Your turn:

10. Subtract $\left(19x^2+5\right)$ from

$\left(81x^2+10\right)$.

	Answer	Text Ref	Video Ref		Answer	Text Ref	Video Ref
1	2, trinomial	Ex 2a, p. 319		6	$-1.1y^2+4.8$		Sec 5.2, 6/11
2	6, trinomial		Sec 5.2, 2/11	7	$-4x^2+6x+2$	Ex 9, p. 323	
3	17	Ex 4b, p. 320		8	$-3x^2+10$		Sec 5.2, 7/11
4	a. -2 b. 4		Sec 5.2, 3/11	9	$-8y^2-4y+17$	Ex 14, p. 324	
5	$13x^2-2$	Ex 6d, pp. 321–322		10	$62x^2+5$		Sec 5.2, 11/11

☐ **Next, insert your homework.** Make sure you attempt all exercises asked of you and show all work, as in the exercises above. Check your answers if possible. Clearly mark any exercises you were unable to correctly complete so that you may ask questions later. DO NOT ERASE YOUR INCORRECT WORK. THIS IS HOW WE UNDERSTAND AND EXPLAIN TO YOU YOUR ERRORS.

Section 5.3 Multiplying Polynomials

Before Class:

☐ Read the objectives on page 328.

☐ Read the **Helpful Hint** box on page 331.

☐ Complete the exercises:

1. To multiply monomials such as $4x^2$ and $-2x^3$, use the _____

   and _____ properties.

2. To multiply a monomial by a binomial, use the _____ property.

During Class:

☐ **Write your class notes.** Neatly write down **all** examples shown as well as key terms or phrases with definitions. If not applicable or if you were absent, watch the Lecture Series (DVD) for this section and do the same (write down the examples shown as well as key terms or phrases). Insert more paper as needed.

Class Notes/Examples	**Your Notes**

Answers: **1)** associative, commutative **2)** distributive

Section 5.3 Multiplying Polynomials

Class Notes (continued)	**Your Notes**

(Insert additional paper as needed.)

Section 5.3 Multiplying Polynomials

Practice:

☐ Complete the Vocabulary, Readiness & Video Check on page 332.

☐ Next, complete any incomplete exercises below. Check and correct your work using the answers and references at the end of this section.

Review this example:

1. Multiply: $\left(-\dfrac{1}{3}x^5\right)\left(-\dfrac{2}{9}x\right)$

$$\left(-\frac{1}{3}x^5\right)\left(-\frac{2}{9}x\right)=\left(-\frac{1}{3}\cdot-\frac{2}{9}\right)\cdot\left(x^5\cdot x\right)$$
$$=\boxed{\frac{2}{27}x^6}$$

Your turn:

2. Multiply: $\left(-\dfrac{1}{3}y^2\right)\left(\dfrac{2}{5}y\right)$

Review this example:

3. Use the distributive property to find each product.

 a. $5x\left(2x^3+6\right)$ b. $-3x^2\left(5x^2+6x-1\right)$

 a. $5x\left(2x^3+6\right)=5x\left(2x^3\right)+5x\left(6\right)$
$$=\boxed{10x^4+30x}$$

 b. $-3x^2\left(5x^2+6x-1\right)$
$$=\left(-3x^2\right)(5x)+\left(-3x^2\right)(6x)+\left(-3x^2\right)(-1)$$
$$=\boxed{-15x^4-18x^3+3x^2}$$

Your turn:

4. Use the distributive property to find the product.

 $-y\left(4x^3-7x^2y+xy^2+3y^3\right)$

Review this example:

5. Multiply: $(3x+2)(2x-5)$

Multiply each term of the first binomial by each term of the second.
$$(3x+2)(2x-5)=3x(2x)+3x(-5)+2(2x)+2(-5)$$
$$=6x^2-15x+4x-10$$
$$=\boxed{6x^2-11x-10}$$

Your turn:

6. Multiply: $(a+7)(a-2)$

Section 5.3 Multiplying Polynomials

Review this example:

7. Multiply $(t+2)$ by $(3t^2-4t+2)$.

Multiply each term of the first polynomial by each term of the second.

$(t+2)(3t^2-4t+2)$

$=t(3t^2)+t(-4t)+t(2)+2(3t^2)+2(-4t)+2(2)$

$=3t^3-4t^2+2t+6t^2-8t+4$

$=\boxed{3t^3+2t^2-6t+4}$

Your turn:

8. Multiply: $(x+5)(x^3-3x+4)$

Review this example:

9. Multiply $(2x^3-3x+4)(x^2+1)$. Use a vertical format.

$$2x^3-3x+4$$
$$\underline{\times \qquad\qquad x^2+1}$$
$$2x^3\qquad -3x+4$$
$$\underline{2x^5-3x^3+4x^2}$$
$$\boxed{2x^5-x^3+4x^2-3x+4}$$

Your turn:

10. Multiply $(5x+1)(2x^2+4x-1)$. Use a vertical format.

	Answer	Text Ref	Video Ref		Answer	Text Ref	Video Ref
1	$\dfrac{2}{27}x^6$	Ex 3, p. 329		6	$a^2+5a-14$		Sec 5.3, 4/7
2	$-\dfrac{2}{15}y^3$		Sec 5.3, 1/7	7	$3t^3+2t^2-6t+4$	Ex 7, p. 330	
3	a. $10x^4+30x$ b. $-15x^4-18x^3+3x^2$	Ex 4, p. 329		8	$x^4+5x^3-3x^2-11x+20$		Sec 5.3, 6/7
4	$-4x^3y+7x^2y^2-xy^3-3y^4$		Sec 5.3, 3/7	9	$2x^5-x^3+4x^2-3x+4$	Ex 10, p. 331	
5	$6x^2-11x-10$	Ex 5, p. 330		10	$10x^3+22x^2-x-1$		Sec 5.3, 7/7

☐ **Next, insert your homework.** Make sure you attempt all exercises asked of you and show all work, as in the exercises above. Check your answers if possible. Clearly mark any exercises you were unable to correctly complete so that you may ask questions later. DO NOT ERASE YOUR INCORRECT WORK. THIS IS HOW WE UNDERSTAND AND EXPLAIN TO YOU YOUR ERRORS.

Before Class:

☐　Read the objectives on page 335.

☐　Read the **Helpful Hint** boxes on pages 335, 337, and 339.

☐　Complete the exercises:

1.　What is the FOIL method?

2.　What do the letters in FOIL represent?

3.　Name two products that are considered "special" products.

During Class:

☐　**Write your class notes.** Neatly write down **all** examples shown as well as key terms or phrases with definitions. If not applicable or if you were absent, watch the Lecture Series (DVD) for this section and do the same (write down the examples shown as well as key terms or phrases). Insert more paper as needed.

Class Notes/Examples	Your Notes

Answers:　**1)** a special order for multiplying binomials　**2)** F stands for the product of the first terms. O stands for the product of the outer terms. I stands for the product of the inner terms. L stands for the product of the last terms.　**3)** Answers may vary.

Section 5.4 Special Products

Class Notes (continued)	**Your Notes**

(Insert additional paper as needed.)

Practice:

☐   Complete the Vocabulary, Readiness & Video Check on page 339.

☐   Next, complete any incomplete exercises below. Check and correct your work using the answers and references at the end of this section.

Review this example:	Your turn:
1. Multiply $(5x-7)(x-2)$ by the FOIL method. $(5x-7)(x-2)$ $\qquad$ F $\qquad$ O $\qquad$ I $\qquad$ L $=5x(x)+5x(-2)+(-7)(x)+(-7)(-2)$ $=5x^2-10x-7x+14$ $=\boxed{5x^2-17x+14}$	**2.** Multiply using the FOIL method. $(x+3)(x+4)$
Review this example: **3.** Use a special product to square the binomial. $(2x+5)^2$ $(2x+5)^2=(2x)^2+2(2x)(5)+5^2$ $\qquad=\boxed{4x^2+20x+25}$	**Your turn:** **4.** Multiply: $(5x+9)^2$
Review this example: **5.** Use a special product to square the binomial. $\left(x^2-7y\right)^2$ $\left(x^2-7y\right)^2=\left(x^2\right)^2-2\left(x^2\right)(7y)+(7y)^2$ $\qquad=\boxed{x^4-14x^2y+49y^2}$	**Your turn:** **6.** Multiply: $(2x-1)^2$
Review this example: **7.** Use a special product to multiply $(6t+7)(6t-7)$. $(6t+7)(6t-7)=(6t)^2-7^2=\boxed{36t^2-49}$	**Your turn:** **8.** Multiply: $(9x+y)(9x-y)$

Section 5.4 Special Products

Review this example:

9. Use a special product to multiply, if possible.

$$(x-5)(3x+4)$$

$$(x-5)(3x+4) = 3x^2 + 4x - 15x - 20$$
$$= 3x^2 - 11x - 20$$

Your turn:

10. Multiply: $(3b+7)(2b-5)$

	Answer	Text Ref	Video Ref		Answer	Text Ref	Video Ref
1	$5x^2 - 17x + 14$	Ex 2, p. 335		6	$4x^2 - 4x + 1$		Sec 5.4, 4/10
2	$x^2 + 7x + 12$		Sec 5.4, 1/10	7	$36t^2 - 49$	Ex 6b, p. 338	
3	$4x^2 + 20x + 25$	Ex 5c, p. 337		8	$81x^2 - y^2$		Sec 5.4, 5/10
4	$25x^2 + 90x + 81$		Sec 5.4, 3/10	9	$3x^2 - 11x - 20$	Ex 7a, p. 338	
5	$x^4 - 14x^2 y + 49y^2$	Ex 5d, p. 337		10	$6b^2 - b - 35$		Sec 5.4, 2/10

☐   **Next, insert your homework.** Make sure you attempt all exercises asked of you and show all work, as in the exercises above. Check your answers if possible. Clearly mark any exercises you were unable to correctly complete so that you may ask questions later. DO NOT ERASE YOUR INCORRECT WORK. THIS IS HOW WE UNDERSTAND AND EXPLAIN TO YOU YOUR ERRORS.

Section 5.5 Negative Exponents and Scientific Notation

Before Class:

☐   Read the objectives on page 342.

☐   Read the **Helpful Hint** boxes on page 343.

☐   Complete the exercises:

1.   A positive number is written in scientific notation if it is written as the product of a

    number a, where _____ a _____, and an integer power r of 10.

2.   To write a scientific notation number in standard form, move the decimal point to the

    _____ if the exponent on 10 is positive; move the decimal

    point to the _____ if the exponent on 10 is negative.

During Class:

☐   **Write your class notes.** Neatly write down **all** examples shown as well as key terms or phrases with definitions. If not applicable or if you were absent, watch the Lecture Series (DVD) for this section and do the same (write down the examples shown as well as key terms or phrases). Insert more paper as needed.

Class Notes/Examples	Your Notes

Answers: **1)** $1 \leq$, < 10 **2)** right, left

Section 5.5 Negative Exponents and Scientific Notation

Class Notes (continued)	Your Notes

(Insert additional paper as needed.)

Section 5.5 Negative Exponents and Scientific Notation

Practice:

☐   Complete the Vocabulary, Readiness & Video Check on page 348.

☐   Next, complete any incomplete exercises below. Check and correct your work using the
      answers and references at the end of this section.

Review this example:	**Your turn:**
1. Simplify $\dfrac{y}{y^{-2}}$.	**2.** Simplify $\dfrac{r}{r^{-3}r^{-2}}$. Write the result using positive exponents only.

$$\frac{y}{y^{-2}} = \frac{y^{1}}{y^{-2}} = y^{1-(-2)} = \boxed{y^{3}}$$

Review this example:	**Your turn:**
3. Simplify $\left(\dfrac{-2x^{3}y}{xy^{-1}}\right)^{3}$. Write the result using positive exponents only.	**4.** Simplify $\dfrac{\left(-2xy^{-3}\right)^{-3}}{\left(xy^{-1}\right)^{-1}}$. Write the result using positive exponents only.

$$\left(\frac{-2x^{3}y}{xy^{-1}}\right)^{3} = \frac{(-2)^{3}x^{9}y^{3}}{x^{3}y^{-3}} = \frac{-8x^{9}y^{3}}{x^{3}y^{-3}} = -8x^{9-3}y^{3-(-3)}$$
$$= \boxed{-8x^{6}y^{6}}$$

Review this example:	**Your turn:**

5. Write each number in scientific notation.
 	
  a. 367,000,000   b. 0.000003
	6. Write each number in scientific notation.
	
	  a. 78,000

a. Step 1. Move the decimal point until the number
           is between 1 and 10.
    Step 2. The decimal point is moved 8 places,
           and the original number is 10 or greater,
           so the count is positive 8.
    Step 3. $367,000,000 = \boxed{3.67 \times 10^{8}}$.

b. 0.00000167

b. Step 1. Move the decimal point until the number
           is between 1 and 10.
    Step 2. The decimal point is moved 6 places,
           and the original number is less than 1, so
           the count is -6.
    Step 3. $0.000003 = \boxed{3.0 \times 10^{-6}}$.

Section 5.5 Negative Exponents and Scientific Notation

Review this example:

7. Write 1.02×10^5 in standard notation, without exponents.

Move the decimal point 5 places to the right.

$1.02 \times 10^5 = \boxed{102,000}$

Your turn:

8. Write 3.3×10^{-2} in standard notation.

Review this example:

9. Perform the indicated operation. Write the result in standard decimal notation.

$$\frac{12 \times 10^2}{6 \times 10^{-3}}$$

$$\frac{12 \times 10^2}{6 \times 10^{-3}} = \frac{12}{6} \times 10^{2-(-3)} = 2 \times 10^5 = \boxed{200,000}$$

Your turn:

10. Evaluate the expression using exponential rules. Write the result in standard notation.

$$\frac{1.4 \times 10^{-2}}{7 \times 10^{-8}}$$

	Answer	Text Ref	Video Ref		Answer	Text Ref	Video Ref
1	y^3	Ex 3a, p. 344		6	a. 7.8×10^4 b. 1.67×10^{-6}		Sec 5.5, 9–10/13
2	r^6		Sec 5.5, 7/13	7	102,000	Ex 6a, p. 347	
3	$-8x^6 y^6$	Ex 4e, p. 345		8	0.033		Sec 5.5, 11/13
4	$-\dfrac{y^8}{8x^2}$		Sec 5.5, 8/13	9	200,000	Ex 7b, p. 347	
5	a. 3.67×10^8 b. 3.0×10^{-6}	Ex 5a, b, p. 346		10	200,000		Sec 5.5, 13/13

☐ **Next, insert your homework.** Make sure you attempt all exercises asked of you and show all work, as in the exercises above. Check your answers if possible. Clearly mark any exercises you were unable to correctly complete so that you may ask questions later. DO NOT ERASE YOUR INCORRECT WORK. THIS IS HOW WE UNDERSTAND AND EXPLAIN TO YOU YOUR ERRORS.

Section 5.6 Dividing Polynomials

Before Class:

☐ Read the objectives on page 351.

☐ Read the **Helpful Hint** box on page 352.

☐ Complete the exercises:

1. To divide a polynomial by a monomial, divide each term of the

   _____ by the _____ .

2. To check a division problem, multiply the _____ by the

   _____ and add the _____ to see if

   the result is equal to the _____ .

During Class:

☐ **Write your class notes.** Neatly write down **all** examples shown as well as key terms or phrases with definitions. If not applicable or if you were absent, watch the Lecture Series (DVD) for this section and do the same (write down the examples shown as well as key terms or phrases). Insert more paper as needed.

Class Notes/Examples	Your Notes

Answers: **1)** polynomial, monomial   **2)** divisor, quotient, remainder, dividend

Section 5.6 Dividing Polynomials

Class Notes (continued)

Your Notes

(Insert additional paper as needed.)

Practice:

☐   Complete the Vocabulary, Readiness & Video Check on page 355.

☐   Next, complete any incomplete exercises below. Check and correct your work using the answers and references at the end of this section.

Review this example:

1. Divide: $\dfrac{9x^5 - 12x^2 + 3x}{3x^2}$

$$\frac{9x^5 - 12x^2 + 3x}{3x^2} = \frac{9x^5}{3x^2} - \frac{12x^2}{3x^2} + \frac{3x}{3x^2}$$

$$= 3x^3 - 4 + \frac{1}{x}$$

Check:

$$3x^2\left(3x^3 - 4 + \frac{1}{x}\right) = 3x^2\left(3x^3\right) - 3x^2\left(4\right) + 3x^2\left(\frac{1}{x}\right)$$

$$= 9x^5 - 12x^2 + 3x$$

Your turn:

2. Divide: $\dfrac{-9x^5 + 3x^4 - 12}{3x^3}$

Review this example:

3. Divide $6x^2 + 10x - 5$ by $3x - 1$ using long division.

$$
\begin{array}{r}
2x + 4 \\
3x - 1 \overline{\smash{)}\, 6x^2 + 10x - 5} \\
\underline{{}^-\,6x^2 \not{+}\, {}^+2x} \\
12x - 5 \\
\underline{{}^-\,12x \not{+}\, {}^+4} \\
-1
\end{array}
$$

Thus $\left(6x^2 + 10x - 5\right)$ divided by $\left(3x - 1\right)$ is $\left(2x + 4\right)$ with a remainder of -1. This can be

written as $\dfrac{6x^2 + 10x - 5}{3x - 1} = 2x + 4 + \dfrac{-1}{3x - 1}$.

Check:

$$(3x - 1)(2x + 4) + (-1) = \left(6x^2 + 12x - 2x - 4\right) - 1$$

$$= 6x^2 + 10x - 5$$

The quotient checks.

Your turn:

4. Find the quotient using long division.

$$\frac{x^2 + 4x + 3}{x + 3}$$

Section 5.6 Dividing Polynomials

Review this example:

5. Divide: $\dfrac{2x^4 - x^3 + 3x^2 + x - 1}{x^2 + 1}$

Rewrite the divisor $x^2 + 1$ as $x^2 + 0x + 1$.

$$
\begin{array}{r}
2x^2 - x + 1 \\
x^2 + 0x + 1 \,\overline{)\, 2x^4 - x^3 + 3x^2 + x - 1} \\
\underline{2x^4 + 0x^3 + 2x^2} \\
-x^3 + x^2 + x \\
\underline{+x^3 + 0x^2 + x} \\
x^2 + 2x - 1 \\
\underline{x^2 + 0x + 1} \\
2x - 2
\end{array}
$$

Thus,

$\dfrac{2x^4 - x^3 + 3x^2 + x - 1}{x^2 + 1} = \boxed{2x^2 - x + 1 + \dfrac{2x - 2}{x^2 + 1}}.$

Your turn:

6. Divide $\dfrac{2b^3 + 9b^2 + 6b - 4}{b + 4}$ using long division.

	Answer	Text Ref	Video Ref		Answer	Text Ref	Video Ref
1	$3x^3 - 4 + \dfrac{1}{x}$	Ex 2, p. 352		**4**	$x + 1$		Sec 5.6, 3/5
2	$-3x^2 + x - \dfrac{4}{x^3}$		Sec 5.6, 2/5	**5**	$2x^2 - x + 1 + \dfrac{2x-2}{x^2+1}$	Ex 7, p. 354	
3	$2x + 4 + \dfrac{-1}{3x-1}$	Ex 5, p. 353		**6**	$2b^2 + b + 2 - \dfrac{12}{b+4}$		Sec 5.6, 4/5

☐ **Next, insert your homework.** Make sure you attempt all exercises asked of you and show all work, as in the exercises above. Check your answers if possible. Clearly mark any exercises you were unable to correctly complete so that you may ask questions later. DO NOT ERASE YOUR INCORRECT WORK. THIS IS HOW WE UNDERSTAND AND EXPLAIN TO YOU YOUR ERRORS.

Preparing for the Chapter 5 Test

Start preparing for your Chapter 5 Test as soon as possible. Pay careful attention to any instructor discussion about this test, especially discussion on what sections you will be responsible for, etc.

☐   Work the Chapter 5 Vocabulary Check on page 358.

☐   Read your Class Notes/Examples for each section covered on your Chapter 5 Test. Look for any unresolved questions you may have.

☐   Complete as many of the Chapter 5 Review exercises as possible (page 361). Remember, the odd answers are in the back of your text.

☐   **Most important:** Place yourself in "test" conditions (see below) and work the Chapter 5 Test (page 363) as a practice test the day before your actual test. To honestly assess how you are doing, try the following:

- Work on a few blank sheets of paper.
- Give yourself the same amount of time you will be given for your actual test.
- Complete this Chapter 5 Practice Test without using your notes or your text.
- If you have any time left after completing this practice test, check your work and try to find any errors on your own.
- Once done, use the back of your book to check ALL answers.
- Try to correct any errors on your own.
- Use the Chapter Test Prep Video (CTPV) to correct any errors you were unable to correct on your own. You can find these videos in the Interactive DVD Lecture Series, in MyMathLab, and on YouTube. Search Martin-Gay Beginning Algebra and click "Channels."

I wish you the best of luck….Elayn Martin-Gay

Section 6.1 The Greatest Common Factor and Factoring by Grouping

Before Class:

☐   Read the objectives on page 367.

☐   Read the **Helpful Hint** boxes on pages 369, 370, 371, 372, and 373.

☐   Complete the exercises:

    1.   What is a prime number?

    2.   The first step in factoring a polynomial is to find the _____ of its terms.

    3.   If a polynomial has four terms we can try factoring by _____ .

During Class:

☐   **Write your class notes.** Neatly write down **all** examples shown as well as key terms or phrases with definitions. If not applicable or if you were absent, watch the Lecture Series (DVD) for this section and do the same (write down the examples shown as well as key terms or phrases). Insert more paper as needed.

Class Notes/Examples	Your Notes

Answers: **1)** a whole number other than 1 whose only factors are 1 and itself   **2)** greatest common factor (GCF)   **3)** grouping

Section 6.1 The Greatest Common Factor and Factoring by Grouping

Class Notes (continued)	Your Notes

(Insert additional paper as needed.)

Section 6.1 The Greatest Common Factor and Factoring by Grouping

Practice:

☐   Complete the Vocabulary, Readiness & Video Check on page 373.

☐   Next, complete any incomplete exercises below. Check and correct your work using the answers and references at the end of this section.

Review this example: **1.** Find the GCF of the terms $6x^2, 10x^3$, and $-8x$. $6x^2 = 2 \cdot 3 \cdot x^2$ $10x^3 = 2 \cdot 5 \cdot x^3$ $-8x = -1 \cdot 2 \cdot 2 \cdot 2 \cdot x^1$ GCF $= 2 \cdot x^1$ or $\boxed{2x}$	**Your turn:** **2.** Find the GCF of the terms $12y^4$ and $20y^3$.
Review this example: **3.** Factor the polynomial $6t + 18$ by factoring out the GCF. The GCF of the terms $6t$ and 18 is 6. $6t + 18 = 6 \cdot t + 6 \cdot 3 = \boxed{6(t+3)}$ Check by multiplying 6 and $(t+3)$. $6(t+3) = 6 \cdot t + 6 \cdot 3 = 6t + 18$, the original polynomial.	**Your turn:** **4.** Factor out the GCF from the polynomial $30x - 15$.
Review this example: **5.** Factor: $5(x+3) + y(x+3)$ The binomial $(x+3)$ is the greatest common factor. Use the distributive property to factor out $(x+3)$. $5(x+3) + y(x+3) = \boxed{(x+3)(5+y)}$	**Your turn:** **6.** Factor out the GCF from the polynomial $y(x^2+2) + 3(x^2+2)$.
Review this example: **7.** Factor $xy + 2x + 3y + 6$ by grouping. Check by multiplying. The GCF of the first two terms is x, and the GCF of the last two terms is 3. *(solution continued on the next page)*	**Your turn:** **8.** Factor by grouping: $5xy - 15x - 6y + 18$

171

Section 6.1 The Greatest Common Factor and Factoring by Grouping

$$xy + 2x + 3y + 6 = (xy + 2x) + (3y + 6)$$
$$= x(y + 2) + 3(y + 2)$$
$$= \boxed{(y + 2)(x + 3)}$$

Check: $(y + 2)(x + 3) = xy + 2x + 3y + 6$, the original polynomial.

Review this example:	Your turn:
9. Factor by grouping: $3x^2 + 4xy - 3x - 4y$	**10.** Factor: $6a^2 + 9ab^2 + 6ab + 9b^3$

$$3x^2 + 4xy - 3x - 4y = (3x^2 + 4xy) + (-3x - 4y)$$
$$= x(3x + 4y) - 1(3x + 4y)$$
$$= \boxed{(3x + 4y)(x - 1)}$$

	Answer	Text Ref	Video Ref		Answer	Text Ref	Video Ref
1	$2x$	Ex 3a, p. 369		6	$(x^2 + 2)(y + 3)$		Sec 6.1, 6/8
2	$4y^3$		Sec 6.1, 3/8	7	$(y + 2)(x + 3)$	Ex 11, p. 371	
3	$6(t + 3)$	Ex 4a, p. 370		8	$(5x - 6)(y - 3)$		Sec 6.1, 7/8
4	$15(2x - 1)$		Sec 6.1, 4/8	9	$(3x + 4y)(x - 1)$	Ex 13, p. 372	
5	$(x + 3)(5 + y)$	Ex 9, p. 371		10	$3(2a + 3b^2)(a + b)$		Sec 6.1, 8/8

☐   **Next, insert your homework.** Make sure you attempt all exercises asked of you and show all work, as in the exercises above. Check your answers if possible. Clearly mark any exercises you were unable to correctly complete so that you may ask questions later. DO NOT ERASE YOUR INCORRECT WORK. THIS IS HOW WE UNDERSTAND AND EXPLAIN TO YOU YOUR ERRORS.

Section 6.2 Factoring Trinomials of the Form $x^2 + bx + c$

Before Class:

☐   Read the objectives on page 375.

☐   Read the **Helpful Hint** boxes on pages 376 and 379.

☐   Complete the exercises:

1.   Factoring and _____ are reverse processes.

2.   A polynomial not factorable with integers is called a _____ polynomial.

3.   If the constant term in a trinomial is negative, look for factors of the constant term with

     _____ signs.

During Class:

☐   **Write your class notes.** Neatly write down **all** examples shown as well as key terms or phrases with definitions. If not applicable or if you were absent, watch the Lecture Series (DVD) for this section and do the same (write down the examples shown as well as key terms or phrases). Insert more paper as needed.

Class Notes/Examples	Your Notes

Answers: **1)** multiplying   **2)** prime   **3)** opposite

Section 6.2 Factoring Trinomials of the Form $x^2 + bx + c$

Class Notes (continued)	**Your Notes**

(Insert additional paper as needed.)

Section 6.2 Factoring Trinomials of the Form $x^2 + bx + c$

Practice:

☐   Complete the Vocabulary, Readiness & Video Check on page 380.

☐   Next, complete any incomplete exercises below. Check and correct your work using the answers and references at the end of this section.

Review this example:	**Your turn:**
1. Factor: $x^2 - 12x + 35$	**2.** Factor the trinomial completely.

Review this example:

1. Factor: $x^2 - 12x + 35$

Look for two numbers whose product is 35 and whose sum is -12. Since our numbers must have a positive product and a negative sum, we look at pairs of negative factors of 35 only.

Negative Factors of 35	Sum of Factors
$-1, -35$	-36
$-5, -7$	-12

Thus, $x^2 - 12x + 35 = (x - 5)(x - 7)$

Your turn:

2. Factor the trinomial completely.

$x^2 - 8x + 15$

Review this example:

3. Factor: $x^2 + 7xy + 6y^2$

The middle term $7xy$ is the same as $7yx$. Thus, $7y$ is the "coefficient" of x.

We then look for two terms whose product is $6y^2$ and whose sum is $7y$. The terms are $6y$ and $1y$ or $6y$ and y because $6y \cdot y = 6y^2$ and $6y + y = 7y$.

Therefore, $x^2 + 7xy + 6y^2 = (x + 6y)(x + y)$

Your turn:

4. Factor the trinomial completely.

$x^2 - 3xy - 4y^2$

Review this example:

5. Factor: $3m^2 - 24m - 60$

First factor out the greatest common factor, 3, from each term.

$3m^2 - 24m - 60 = 3(m^2 - 8m - 20)$

Now factor $m^2 - 8m - 20$ by looking for two factors of -20 whose sum is -8. The factors are -10 and 2.

Therefore, the complete factored form is

$3m^2 - 24m - 60 = 3(m + 2)(m - 10)$

Your turn:

6. Factor the trinomial completely. Don't forget to factor out the GCF first.

$3x^2 + 9x - 30$

Section 6.2 Factoring Trinomials of the Form $x^2 + bx + c$

Review this example:

7. Factor: $2x^4 - 26x^3 + 84x^2$

$$2x^4 - 26x^3 + 84x^2 = 2x^2\left(x^2 - 13x + 42\right)$$
$$= \boxed{2x^2(x-6)(x-7)}$$

Your turn:

8. Factor the trinomial completely. Don't forget to factor out the GCF first.

$$5x^3y - 25x^2y^2 - 120xy^3$$

	Answer	Text Ref	Video Ref		Answer	Text Ref	Video Ref
1	$(x-5)(x-7)$	Ex 2, p. 377		**5**	$3(m+2)(m-10)$	Ex 9, p. 379	
2	$(x-5)(x-3)$		Sec 6.2, 2/6	**6**	$3(x+5)(x-2)$		Sec 6.2, 5/6
3	$(x+6y)(x+y)$	Ex 6, p. 378		**7**	$2x^2(x-6)(x-7)$	Ex 10, p. 379	
4	$(x-4y)(x+y)$		Sec 6.2, 4/6	**8**	$5xy(x-8y)(x+3y)$		Sec 6.2, 6/6

☐   **Next, insert your homework.** Make sure you attempt all exercises asked of you and show all work, as in the exercises above. Check your answers if possible. Clearly mark any exercises you were unable to correctly complete so that you may ask questions later. DO NOT ERASE YOUR INCORRECT WORK. THIS IS HOW WE UNDERSTAND AND EXPLAIN TO YOU YOUR ERRORS.

Section 6.3 Factoring Trinomials of the Form $ax^2 + bx + c$ and Perfect Square Trinomials

Before Class:

☐   Read the objectives on page 382.

☐   Read the **Helpful Hint** boxes on pages 384, 385, 386, and 387.

☐   Complete the exercises:

      1.   If the terms of a trinomial have no common factor other than 1, will the terms of its binomial factors have any common factors?

      2.   What is a perfect square trinomial?

During Class:

☐   **Write your class notes.** Neatly write down **all** examples shown as well as key terms or phrases with definitions. If not applicable or if you were absent, watch the Lecture Series (DVD) for this section and do the same (write down the examples shown as well as key terms or phrases). Insert more paper as needed.

Class Notes/Examples	Your Notes

Answers: **1)** no   **2)** a trinomial that is the square of a binomial

Section 6.3 Factoring Trinomials of the Form $ax^2 + bx + c$ and Perfect Square Trinomials

Class Notes (continued)	**Your Notes**

(Insert additional paper as needed.)

Section 6.3 Factoring Trinomials of the Form $ax^2 + bx + c$ and Perfect Square Trinomials

Practice:

☐   Complete the Vocabulary, Readiness & Video Check on page 388.

☐   Next, complete any incomplete exercises below. Check and correct your work using the
answers and references at the end of this section.

Review this example:	**Your turn:**
1. Factor: $10x^2 - 13xy - 3y^2$	**2.** Factor $4x^2 - 8x - 21$ completely.

Factors of $10x^2$: $10x^2 = 10x \cdot x$ $10x^2 = 2x \cdot 5x$
Factors of $-3y^2$: $-3y^2 = -3y \cdot y$ $-3y^2 = 3y \cdot -y$

Try some combinations of these factors:

$(10x - 3y)(x + y) = 10x^2 + 7xy - 3y^2$
$(x + 3y)(10x - y) = 10x^2 + 29xy - 3y^2$
$(5x + 3y)(2x - y) = 10x^2 + xy - 3y^2$
$(2x - 3y)(5x + y) = 10x^2 - 13xy - 3y^2$ correct

A factored form of $10x^2 - 13xy - 3y^2$ is

$(2x - 3y)(5x + y)$.

Review this example:	**Your turn:**
3. Factor: $3x^4 - 5x^2 - 8$	**4.** Factor $30x^3 + 38x^2 + 12x$ completely.

Factors of $3x^4$: $3x^4 = 3x^2 \cdot x^2$
Factors of -8: $-8 = -2 \cdot 4,\ 2 \cdot -4,\ -1 \cdot 8,\ 1 \cdot -8$

Try combinations of these factors:

$(3x^2 - 2)(x^2 + 4) = 3x^4 + 10x^2 - 8$
$(3x^2 + 4)(x^2 - 2) = 3x^4 - 2x^2 - 8$
$(3x^2 + 8)(x^2 - 1) = 3x^4 + 5x^2 - 8$
$(3x^2 - 8)(x^2 + 1) = 3x^4 - 5x^2 - 8$ correct

A factored form of $3x^4 - 5x^2 - 8$ is

$(3x^2 - 8)(x^2 + 1)$.

Section 6.3 Factoring Trinomials of the Form $ax^2 + bx + c$ and Perfect Square Trinomials

Review this example:

5. Factor: $x^2 + 12x + 36$

Is $x^2 + 12x + 36$ a perfect square trinomial?
1. $x^2 = (x)^2$ and $36 = 6^2$
2. Is $2 \cdot x \cdot 6$ the middle term? Yes, $2 \cdot x \cdot 6 = 12x$.

Thus, $x^2 + 12x + 36$ factors as $(x+6)^2$.

Your turn:

6. Factor: $x^2 + 22x + 121$

Review this example:

7. Factor: $25x^2 + 25xy + 4y^2$

Is $25x^2 + 25xy + 4y^2$ a perfect square trinomial?
1. $25x^2 = (5x)^2$ and $4y^2 = (2y)^2$.
2. Is $2 \cdot 5x \cdot 2y$ the middle term? No,
   $2 \cdot 5x \cdot 2y = 20xy$, not $25xy$.

Therefore, $25x^2 + 25xy + 4y^2$ is not a perfect square trinomial. It is factorable, though. Using earlier techniques, $25x^2 + 25xy + 4y^2$ factors as

$(5x + 4y)(5x + y)$.

Your turn:

8. Factor: $9x^2 - 24xy + 16y^2$

	Answer	Text Ref	Video Ref			Answer	Text Ref	Video Ref
1	$(2x-3y)(5x+y)$	Ex 4, p. 385			5	$(x+6)^2$	Ex 8, p. 387	
2	$(2x-7)(2x+3)$		Sec 6.3, 2/8		6	$(x+11)^2$		Sec 6.3, 7/8
3	$(3x^2-8)(x^2+1)$	Ex 5, p. 385			7	$(5x+4y)(5x+y)$	Ex 9, p. 387	
4	$2x(3x+2)(5x+3)$		Sec 6.3, 3/8		8	$(3x-4y)^2$		Sec 6.3, 8/8

☐   **Next, insert your homework.** Make sure you attempt all exercises asked of you and show all work, as in the exercises above. Check your answers if possible. Clearly mark any exercises you were unable to correctly complete so that you may ask questions later. DO NOT ERASE YOUR INCORRECT WORK. THIS IS HOW WE UNDERSTAND AND EXPLAIN TO YOU YOUR ERRORS.

Section 6.4 Factoring Trinomials of the Form $ax^2 + bx + c$ by Grouping

Before Class:

☐   Read the objective on page 390.

☐   Complete the exercises:

1.   Read the To Factor Trinomials by Grouping box on page 391. After the greatest common factor has been factored out of a trinomial, what should be done to the resulting trinomial $ax^2 + bx + c$?

2.   After completing the step named in exercise 1, what is left to be done to factor the trinomial by grouping?

During Class:

☐   **Write your class notes.** Neatly write down **all** examples shown as well as key terms or phrases with definitions. If not applicable or if you were absent, watch the Lecture Series (DVD) for this section and do the same (write down the examples shown as well as key terms or phrases). Insert more paper as needed.

Class Notes/Examples	Your Notes

Answers: **1)** Find two numbers whose product is $a \cdot c$ and whose sum is b.   **2)** Write the middle term of the trinomial, bx, using the factors found in exercise 1.

Section 6.4 Factoring Trinomials of the Form $ax^2 + bx + c$ by Grouping

Class Notes (continued)

Your Notes

(Insert additional paper as needed.)

Section 6.4 Factoring Trinomials of the Form $ax^2 + bx + c$ by Grouping

Practice:

☐   Complete the Vocabulary, Readiness & Video Check on page 394.

☐   Next, complete any incomplete exercises below. Check and correct your work using the answers and references at the end of this section.

Review this example:	**Your turn:**
1. Factor $3x^2 + 31x + 10$ by grouping.	**2.** Factor by grouping: $21y^2 + 17y + 2$

Step 1. The terms of this trinomial contain no greatest common factor other than 1 or -1.

Step 2. In $3x^2 + 31x + 10$, $a = 3$, $b = 31$, and $c = 10$. Find two numbers whose product is $a \cdot c$ or $3(10) = 30$ and whose sum is b or 31. The numbers are 1 and 30.

Factors of 30	Sum of factors
5, 6	11
3, 10	13
2, 15	17
1, 30	31

Step 3. Write $31x$ as $1x + 30x$ so that
$3x^2 + 31x + 10 = 3x^2 + 1x + 30x + 10$.

Step 4: Factor by grouping.
$$3x^2 + 1x + 30x + 10 = x(3x + 1) + 10(3x + 1)$$
$$= (3x + 1)(x + 10)$$

Review this example:	**Your turn:**
3. Factor $8x^2 - 14x + 5$ by grouping.	**4.** Factor by grouping: $10x^2 - 9x + 2$

Step 1. The terms of this trinomial contain no greatest common factor other than 1.

Step 2. In $8x^2 - 14x + 5$, $a = 8$, $b = -14$, and $c = 5$. Find two numbers whose product is $a \cdot c$ or $8 \cdot 5 = 40$ and whose sum is b or -14. The numbers are -4 and -10.

Factors of 40	Sum of factors
$-40, -1$	-41
$-20, -2$	-22
$-10, -4$	-14

(solution continued on the next page)

Section 6.4 Factoring Trinomials of the Form $ax^2 + bx + c$ by Grouping

Step 3. Write $-14x$ as $-4x - 10x$ so that
$8x^2 - 14x + 5 = 8x^2 - 4x - 10x + 5$.

Step 4. Factor by grouping.
$8x^2 - 4x - 10x + 5 = 4x(2x - 1) - 5(2x - 1)$
$= (2x - 1)(4x - 5)$

Review this example:

5. Factor $18y^4 + 21y^3 - 60y^2$ by grouping.

Step 1. Factor out the greatest common factor, $3y^2$.
$18y^4 + 21y^3 - 60y^2 = 3y^2(6y^2 + 7y - 20)$

Step 2. Notice that $a = 6$, $b = 7$, and $c = -20$ in the resulting trinomial. Find two numbers whose product is $a \cdot c$ or $6(-20) = -120$ and whose sum is 7.

It may help to factor -120 as a product of primes and -1: $-120 = 2 \cdot 2 \cdot 2 \cdot 3 \cdot 5 \cdot (-1)$. Choose pairings of factors until you have two factors whose sum is 7. The numbers are -8 and 15.

Step 3. $6y^2 + 7y - 20 = 6y^2 - 8y + 15y - 20$

Step 4. $6y^2 - 8y + 15y - 20 = 2y(3y - 4) + 5(3y - 4)$
$= (3y - 4)(2y + 5)$

The factored form of $18y^4 + 21y^3 - 60y^2$ is

$3y^2(3y - 4)(2y + 5)$.

Your turn:

6. Factor by grouping:
$12x^3 - 27x^2 - 27x$

	Answer	Text Ref	Video Ref		Answer	Text Ref	Video Ref
1	$(3x+1)(x+10)$	Ex 1, pp. 391–392		4	$(5x-2)(2x-1)$		Sec 6.4, 3/4
2	$(7y+1)(3y+2)$		Sec 6.4, 2/4	5	$3y^2(3y-4)(2y+5)$	Ex 4, p. 393	
3	$(2x-1)(4x-5)$	Ex 2, p.392		6	$3x(x-3)(4x+3)$		Sec 6.4, 4/4

☐   **Next, insert your homework.** Make sure you attempt all exercises asked of you and show all work, as in the exercises above. Check your answers if possible. Clearly mark any exercises you were unable to correctly complete so that you may ask questions later. DO NOT ERASE YOUR INCORRECT WORK. THIS IS HOW WE UNDERSTAND AND EXPLAIN TO YOU YOUR ERRORS.

Before Class:

☐   Read the objectives on page 395.

☐   Read the **Helpful Hint** boxes on pages 396, 397, and 398.

☐   Complete the exercises:

    1. Write the pattern for factoring a difference of squares.

    2. Write the pattern for factoring the sum of two cubes.

    3. Write the pattern for factoring the difference of two cubes.

During Class:

☐   **Write your class notes.** Neatly write down **all** examples shown as well as key terms or phrases with definitions. If not applicable or if you were absent, watch the Lecture Series (DVD) for this section and do the same (write down the examples shown as well as key terms or phrases). Insert more paper as needed.

Class Notes/Examples	Your Notes

Answers: **1)** $a^2 - b^2 = (a+b)(a-b)$ **2)** $a^3 + b^3 = (a+b)(a^2 - ab + b^2)$

3) $a^3 - b^3 = (a-b)(a^2 + ab + b^2)$

Section 6.5 Factoring Binomials

Class Notes (continued)	Your Notes

(Insert additional paper as needed.)

Copyright © 2013 Pearson Education, Inc.

Practice:

☐   Complete the Vocabulary, Readiness & Video Check on page 400.

☐   Next, complete any incomplete exercises below. Check and correct your work using the answers and references at the end of this section.

Review this example:	**Your turn:**
1. Factor the difference of squares.	**2.** Factor completely: $121m^2 - 100n^2$
$25a^2 - 9b^2$	
$25a^2 - 9b^2 = (5a)^2 - (3b)^2 = \boxed{(5a+3b)(5a-3b)}$	

Review this example:	**Your turn:**
3. Factor the binomial: $x^2 + 4$	**4.** Factor completely: $16r^2 + 1$
The binomial $x^2 + 4$ is the sum of two squares since we can write $x^2 + 4$ as $x^2 + 2^2$.	
We might try to factor using $(x+2)(x+2)$ or $(x-2)(x-2)$, but when we multiply to check, we find that neither factoring is correct.	
$x^2 + 4$ is a $\boxed{prime}$ polynomial.	

Review this example:	**Your turn:**
5. Factor the difference of two squares:	**6.** Factor completely: $xy^3 - 9xyz^2$
$\quad 4x^3 - 49x$	
$4x^3 - 49x = x(4x^2 - 49)$	
$\quad\quad = x\left[(2x)^2 - 7^2\right]$	
$\quad\quad = \boxed{x(2x+7)(2x-7)}$	

Review this example:	**Your turn:**
7. Factor: $-49x^2 + 16$	**8.** Factor completely: $49 - \dfrac{9}{25}m^2$
Factor as is, or rearrange terms.	
Factor as is: $-49x^2 + 16 = -1(49x^2 - 16)$	
$\quad\quad\quad\quad = -1(7x+4)(7x-4)$	
Rewrite binomial: $-49x^2 + 16 = 16 - 49x^2$	
$\quad\quad\quad\quad = \boxed{(4+7x)(4-7x)}$	
Both factorizations are correct and equal.	

Section 6.5 Factoring Binomials

Review this example:

9. Factor: $x^3 + 8$

First, write the binomial in the form $a^3 + b^3$.
$x^3 + 8 = x^3 + 2^3$

Replace a with x and b with 2 in the formula for the sum of two cubes.

$x^3 + 2^3 = (x+2)\left(x^2 - (x)(2) + 2^2\right)$

$= \boxed{(x+2)\left(x^2 - 2x + 4\right)}$

Your turn:

10. Factor the sum of two cubes:
$x^3 + 125$

Review this example:

11. Factor: $y^3 - 27$

$y^3 - 27 = y^3 - 3^3$

$= (y-3)\left(y^2 + (y)(3) + 3^2\right)$

$= \boxed{(y-3)\left(y^2 + 3y + 9\right)}$

Your turn:

12. Factor the difference of two cubes:
$x^3 y^3 - 64$

	Answer	Text Ref	Video Ref		Answer	Text Ref	Video Ref
1	$(5a+3b)(5a-3b)$	Ex 2b, p. 396		7	$(4+7x)(4-7x)$	Ex 7, p. 397	
2	$(11m+10n)(11m-10n)$		Sec 6.5, 2/8	8	$\left(7+\dfrac{3}{5}m\right)\left(7-\dfrac{3}{5}m\right)$		Sec 6.5, 5/8
3	prime	Ex 4b, pp. 396–397		9	$(x+2)\left(x^2 - 2x + 4\right)$	Ex 8, p. 398	
4	prime		Sec 6.5, 3/8	10	$(x+5)\left(x^2 - 5x + 25\right)$		Sec 6.5, 6/8
5	$x(2x+7)(2x-7)$	Ex 5, p. 397		11	$(y-3)\left(y^2 + 3y + 9\right)$	Ex 9, p. 398	
6	$xy(y+3z)(y-3z)$		Sec 6.5, 4/8	12	$(xy-4)\left(x^2 y^2 + 4xy + 16\right)$		Sec 6.5, 7/8

☐ **Next, insert your homework.** Make sure you attempt all exercises asked of you and show all work, as in the exercises above. Check your answers if possible. Clearly mark any exercises you were unable to correctly complete so that you may ask questions later. DO NOT ERASE YOUR INCORRECT WORK. THIS IS HOW WE UNDERSTAND AND EXPLAIN TO YOU YOUR ERRORS.

Section 6.6 Solving Quadratic Equations by Factoring

Before Class:

☐   Read the objectives on page 405.

☐   Read the **Helpful Hint** boxes on pages 406 and 409.

☐   Complete the exercises:

1.   According to the zero factor property, if a and b are real numbers and if $ab = 0$, then what must be true about a or b?

2.   To use the zero factor property to solve an equation, one side of the equation must be

      written as a _____ and the other side must be equal to

      _____ .

3.   The x-intercepts of the graph of $y = ax^2 + bx + c$ are the real number solutions of what equation?

During Class:

☐   **Write your class notes.** Neatly write down **all** examples shown as well as key terms or phrases with definitions. If not applicable or if you were absent, watch the Lecture Series (DVD) for this section and do the same (write down the examples shown as well as key terms or phrases). Insert more paper as needed.

Class Notes/Examples	**Your Notes**

Answers: **1)** $a = 0$ or $b = 0$ **2)** product, 0   **3)** $ax^2 + bx + c = 0$

Section 6.6 Solving Quadratic Equations by Factoring

Class Notes (continued)	**Your Notes**

(Insert additional paper as needed.)

Section 6.6 Solving Quadratic Equations by Factoring

Practice:

☐   Complete the Vocabulary, Readiness & Video Check on page 412.

☐   Next, complete any incomplete exercises below. Check and correct your work using the answers and references at the end of this section.

Review this example:

1. Solve: $(x-3)(x+1)=0$

If this equation is true, then either $x-3=0$ or $x+1=0$. When solving these two linear equations, $x=3$ or $x=-1$. Thus, 3 and -1 are both solutions of $(x-3)(x+1)=0$.

To check, replace x with 3 and x with -1 in the original equation.

Let $x=3$.

$(x-3)(x+1)=0$

$(3-3)(3+1)\overset{?}{=}0$

$0(4)=0$ True

Let $x=-1$.

$(x-3)(x+1)=0$

$(-1-3)(-1+1)\overset{?}{=}0$

$(-4)(0)=0$ True

The solutions are $\boxed{3 \text{ and } -1.}$

Your turn:

2. Solve: $(2x+3)(4x-5)=0$

Review this example:

3. Solve: $x^2-9x-22=0$

One side of the equation is zero. Factor the polynomial: $x^2-9x-22=0$

$(x-11)(x+2)=0$

Apply the zero factor property.

$x-11=0$ or $x+2=0$

$x=11$ $x=-2$

Check: Let $x=11$.

$x^2-9x-22=0$

$11^2-9\cdot11-22\overset{?}{=}0$

$121-99-22\overset{?}{=}0$

$22-22\overset{?}{=}0$

True $0=0$

Let $x=-2$.

$x^2-9x-22=0$

$(-2)^2-9(-2)-22\overset{?}{=}0$

$4+18-22\overset{?}{=}0$

$22-22\overset{?}{=}0$

True $0=0$

The solutions are $\boxed{11 \text{ and } -2.}$

Your turn:

4. Solve: $x^2+2x-8=0$

Section 6.6 Solving Quadratic Equations by Factoring

Review this example:

5. Solve: $x(2x-7)=4$

$x(2x-7)=4$ Write the equation in

$2x^2-7x=4$ standard form, then factor.

$2x^2-7x-4=0$

$(2x+1)(x-4)=0$

$2x+1=0$ or $x-4=0$

$2x=-1$ $x=4$

$x=-\dfrac{1}{2}$

Check the solutions in the original equation. The solutions are $-\dfrac{1}{2}$ and $4.$

Your turn:

6. Solve: $x(3x-1)=14$

Review this example:

7. Find the x-intercepts of the graph of $y=x^2-5x+4$.

Let $y=0$ and solve for x.

$y=x^2-5x+4$

$0=x^2-5x+4$

$0=(x-1)(x-4)$

$x-1=0$ or $x-4=0$

$x=1$ $x=4$

The x-intercepts of the graph of $y=x^2-5x+4$ are $(1,0)$ and $(4,0)$.

Your turn:

8. Find the x-intercepts of the graph of $y=2x^2+11x-6$.

	Answer	Text Ref	Video Ref			Answer	Text Ref	Video Ref
1	$3,-1$	Ex 1, p. 406		**5**	$-\dfrac{1}{2},4$	Ex 6, p. 409	
2	$-\dfrac{3}{2},\dfrac{5}{4}$		Sec 6.6, 2/5	**6**	$\dfrac{7}{3},-2$		Sec 6.6, 3/5
3	$11,-2$	Ex 4 pp. 407–408		**7**	$(1,0),(4,0)$	Ex 11, p. 411	
4	$-4,2$		Sec 6.6, 1/5	**8**	$\left(\dfrac{1}{2},0\right),(-6,0)$		Sec 6.6, 5/5

☐   **Next, insert your homework.** Make sure you attempt all exercises asked of you and show all work, as in the exercises above. Check your answers if possible. Clearly mark any exercises you were unable to correctly complete so that you may ask questions later. DO NOT ERASE YOUR INCORRECT WORK. THIS IS HOW WE UNDERSTAND AND EXPLAIN TO YOU YOUR ERRORS.

Section 6.7 Quadratic Equations and Problem Solving

Before Class:

☐   Read the objective on page 414.

☐   Read the **Helpful Hint** box on page 418.

☐   Complete the exercises:

1.   When solving a problem using a quadratic equation that models the problem, what must you keep in mind regarding the solutions of the equation with respect to the solutions of the problem?

2.   The hypotenuse of a right triangle is the side opposite the _____ angle.

During Class:

☐   **Write your class notes.** Neatly write down **all** examples shown as well as key terms or phrases with definitions. If not applicable or if you were absent, watch the Lecture Series (DVD) for this section and do the same (write down the examples shown as well as key terms or phrases). Insert more paper as needed.

Class Notes/Examples	**Your Notes**

Answers: **1)** Answers may vary.   **2)** right

Section 6.7 Quadratic Equations and Problem Solving

Class Notes (continued)	**Your Notes**

(Insert additional paper as needed.)

Section 6.7 Quadratic Equations and Problem Solving

Practice:

☐   Complete the Vocabulary, Readiness & Video Check on page 420.

☐   Next, complete any incomplete exercises below. Check and correct your work using the answers and references at the end of this section.

Review this example:

1. Since the 1940s, one of the top tourist attractions in Acapulco, Mexico, is watching the cliff divers off La Quebrada. The divers' platform is about 144 feet above the sea. These divers must time their descent just right, since they land in the crashing Pacific, in an inlet that is at most $9\frac{1}{2}$ feet deep.

   Neglecting air resistance, the height h in feet of a cliff diver above the ocean after t seconds is given by the quadratic equation $h = -16t^2 + 144$. Find out how long it takes the diver to reach the ocean.

UNDERSTAND. Read and reread the problem. The equation $h = -16t^2 + 144$ models the height of the falling diver at time t.

TRANSLATE. To find out how long it takes the diver to reach the ocean, we want to know the value of t for which $h = 0$.

$0 = -16t^2 + 144$

SOLVE. $0 = -16t^2 + 144$

$\qquad 0 = -16(t^2 - 9)$

$\qquad 0 = -16(t - 3)(t + 3)$

$\qquad t - 3 = 0 \ \text{ or } \ t + 3 = 0$

$\qquad\quad t = 3 \qquad\qquad t = -3$

INTERPRET. Since the time t cannot be negative, the proposed solution is 3 seconds.

Check: Verify that the height of the diver when t is 3 seconds is 0.

$h = -16(3)^2 + 144 = -144 + 144 = 0$.

State: It takes the diver ⟨3 seconds⟩ to reach the ocean.

Your turn:

2. An object is thrown upward from the top of an 80-foot building with an initial velocity of 64 feet per second. The height h of the object after t seconds is given by the quadratic equation $h = -16t^2 + 64t + 80$. When will the object hit the ground?

Section 6.7 Quadratic Equations and Problem Solving

Review this example:

3. The square of a number plus three times the number is 70. Find the number.

UNDERSTAND. Read and reread the problem. Let x = the number.

TRANSLATE. $x^2 + 3x = 70$

SOLVE. $x^2 + 3x = 70$
           $x^2 + 3x - 70 = 0$
           $(x + 10)(x - 7) = 0$
           $x + 10 = 0$ or $x - 7 = 0$
                 $x = -10$ $x = 7$

INTERPRET.

Check: The square of -10 is $(-10)^2$, or 100.

Three times -10 is $3(-10)$ or -30. Then

$100 + (-30) = 70$, the correct sum, so -10 checks.

The square of 7 is 7^2 or 49. Three times 7 is $3(7)$,

or 21. Then $49 + 21 = 70$, the correct sum, so 7 checks.

State: There are two numbers. They are -10 and 7.

Your turn:

4. The sum of a number and its square is 182. Find the number(s).

Review this example:

5. Find two consecutive even integers whose product is 34 more than their sum.

UNDERSTAND. Read and reread the problem. Let x and $x + 2$ be the consecutive even integers.

TRANSLATE. $x(x + 2) = x + (x + 2) + 34$

SOLVE. Now solve the equation.
$$x(x + 2) = x + (x + 2) + 34$$
$$x^2 + 2x = x + x + 2 + 34$$
$$x^2 + 2x = 2x + 36$$
$$x^2 - 36 = 0$$
$$(x + 6)(x - 6) = 0$$

(solution continued on the next page)

Your turn:

6. The product of two consecutive page numbers is 420. Find the page numbers.

Section 6.7 Quadratic Equations and Problem Solving

$$x+6=0 \quad \text{or} \quad x-6=0$$
$$x=-6 \qquad x=6$$

INTERPRET. If $x=-6$, then $x+2=-6+2$, or -4. If $x=6$, then $x+2=6+2$, or 8.

Check: $-6,-4$ $6,8$

$$-6(-4)\overset{?}{=}-6+(-4)+34 \qquad 6(8)\overset{?}{=}6+8+34$$

$$24\overset{?}{=}-10+34 \qquad\qquad 48\overset{?}{=}14+34$$

$$24=24 \quad \text{True} \qquad\qquad 48=48 \quad \text{True}$$

State: The two consecutive even integers are

-6 and -4 or 6 and 8 .

Review this example:	Your turn:
7. Find the lengths of the sides of a right triangle if the lengths can be expressed as three consecutive even integers.	**8.** One leg of a right triangle is 4 millimeters longer than the smaller leg and the hypotenuse is 8 millimeters longer than the smaller leg. Find the lengths of the sides of the triangle.

UNDERSTAND. Read and reread the problem.
Let x, $x + 2$, and $x + 4$ be three consecutive even integers. Since these integers represent lengths of the sides of a right triangle,
x = one leg
$x + 2$ = other leg
$x + 4$ = hypotenuse (longest side)

TRANSLATE. By the Pythagorean theorem,
$$(\text{leg})^2+(\text{leg})^2=(\text{hypotenuse})^2$$
$$(x)^2+(x+2)^2=(x+4)^2$$

SOLVE.
$$x^2+(x+2)^2=(x+4)^2$$
$$x^2+x^2+4x+4=x^2+8x+16$$
$$2x^2+4x+4=x^2+8x+16$$
$$x^2-4x-12=0$$
$$(x-6)(x+2)=0$$
$$x-6=0 \quad \text{or} \quad x+2=0$$
$$x=6 \qquad x=-2$$

(solution continued on the next page)

Section 6.7 Quadratic Equations and Problem Solving

INTERPRET. We discard $x = -2$ since length cannot be negative. If $x = 6$, then $x + 2 = 8$ and $x + 4 = 10$.

Check: Verify that

$$(\text{leg})^2 + (\text{leg})^2 = (\text{hypotenuse})^2$$

$$6^2 + 8^2 \overset{?}{=} 10^2$$

$$36 + 64 \overset{?}{=} 100$$

$$100 = 100$$

State: The sides of the right triangle have lengths

6 units, 8 units, and 10 units.

	Answer	Text Ref	Video Ref		Answer	Text Ref	Video Ref
1	3 sec	Ex 1, p. 415		**5**	−6, −4 or 6, 8	Ex 4, pp. 417–418	
2	5 sec		Sec 6.7, 2/5	**6**	20 and 21		Sec 6.7, 5/5
3	−10 or 7	Ex 2, pp. 415–416		**7**	6 units, 8 units, 10 units	Ex 5, pp. 418–419	
4	−14 or 13		Sec 6.7, 4/5	**8**	12 mm, 16 mm, 20 mm		Sec 6.7, 3/5

☐ **Next, insert your homework.** Make sure you attempt all exercises asked of you and show all work, as in the exercises above. Check your answers if possible. Clearly mark any exercises you were unable to correctly complete so that you may ask questions later. DO NOT ERASE YOUR INCORRECT WORK. THIS IS HOW WE UNDERSTAND AND EXPLAIN TO YOU YOUR ERRORS.

Preparing for the Chapter 6 Test

Start preparing for your Chapter 6 Test as soon as possible. Pay careful attention to any instructor discussion about this test, especially discussion on what sections you will be responsible for, etc.

☐ Work the Chapter 6 Vocabulary Check on page 423.

☐ Read your Class Notes/Examples for each section covered on your Chapter 6 Test. Look for any unresolved questions you may have.

☐ Complete as many of the Chapter 6 Review exercises as possible (page 427). Remember, the odd answers are in the back of your text.

☐ **Most important:** Place yourself in "test" conditions (see below) and work the Chapter 6 Test (page 429) as a practice test the day before your actual test. To honestly assess how you are doing, try the following:

- Work on a few blank sheets of paper.
- Give yourself the same amount of time you will be given for your actual test.
- Complete this Chapter 6 Practice Test without using your notes or your text.
- If you have any time left after completing this practice test, check your work and try to find any errors on your own.
- Once done, use the back of your book to check ALL answers.
- Try to correct any errors on your own.
- Use the Chapter Test Prep Video (CTPV) to correct any errors you were unable to correct on your own. You can find these videos in the Interactive DVD Lecture Series, in MyMathLab, and on YouTube. Search Martin-Gay Beginning Algebra and click "Channels."

I wish you the best of luck....Elayn Martin-Gay

Section 7.1 Simplifying Rational Expressions

Before Class:

☐ Read the objectives on page 433.

☐ Read the **Helpful Hint** boxes on pages 434, 437, and 438.

☐ Complete the exercises:

1. A rational expression is _____ for values that make the denominator 0.

2. To simplify a rational expression, _____ the numerator and

    denominator, then divide out _____ common to the numerator and denominator.

During Class:

☐ **Write your class notes.** Neatly write down **all** examples shown as well as key terms or phrases with definitions. If not applicable or if you were absent, watch the Lecture Series (DVD) for this section and do the same (write down the examples shown as well as key terms or phrases). Insert more paper as needed.

Class Notes/Examples	**Your Notes**

Answers: **1)** undefined   **2)** completely factor, factors

Section 7.1 Simplifying Rational Expressions

Class Notes (continued)	**Your Notes**

(Insert additional paper as needed.)

Copyright © 2013 Pearson Education, Inc.

Section 7.1 Simplifying Rational Expressions

Practice:

☐   Complete the Vocabulary, Readiness & Video Check on page 439.

☐   Next, complete any incomplete exercises below. Check and correct your work using the answers and references at the end of this section.

Review this example:

1. Find the value of $\dfrac{x+4}{2x-3}$ for the given replacement values.

     a. $x=5$ b. $x=-2$

a. Replace each x in the expression with 5 and then simplify.

$$\frac{x+4}{2x-3} = \frac{5+4}{2(5)-3} = \frac{9}{10-3} = \boxed{\frac{9}{7}}$$

b. Replace each x in the expression with -2 and then simplify.

$$\frac{x+4}{2x-3} = \frac{-2+4}{2(-2)-3} = \frac{2}{-7} \text{ or } \boxed{-\frac{2}{7}}$$

Your turn:

2. Evaluate $\dfrac{z}{z^2-5}$ when $z=-5$.

Review this example:

3. Are there any values for x for which $\dfrac{x^2+2}{3x^2-5x+2}$ is undefined?

Set the denominator equal to zero.

$$3x^2-5x+2=0$$
$$(3x-2)(x-1)=0$$
$$3x-2=0 \quad \text{or} \quad x-1=0$$
$$3x=2 \qquad\qquad x=1$$
$$x=\frac{2}{3}$$

When $x=\dfrac{2}{3}$ or $x=1$, the denominator $3x^2-5x+2$ is 0. So the rational expression $\dfrac{x^2+2}{3x^2-5x+2}$ is undefined when $\boxed{x=\dfrac{2}{3} \text{ or } x=1.}$

Your turn:

4. Find values for which $\dfrac{11x^2+1}{x^2-5x-14}$ is undefined.

Section 7.1 Simplifying Rational Expressions

Review this example:

5. Simplify: $\dfrac{x^2+4x+4}{x^2+2x}$

We factor the numerator and denominator and then look for common factors.

$\dfrac{x^2+4x+4}{x^2+2x} = \dfrac{(x+2)(x+2)}{x(x+2)} = \boxed{\dfrac{x+2}{x}}$

Your turn:

6. Simplify: $\dfrac{x^3+7x^2}{x^2+5x-14}$

Review this example:

7. Simplify: $\dfrac{x+9}{x^2-81}$

We factor and then divide out common factors.

$\dfrac{x+9}{x^2-81} = \dfrac{x+9}{(x+9)(x-9)} = \boxed{\dfrac{1}{x-9}}$

Your turn:

8. Simplify: $\dfrac{4-x^2}{x-2}$

Review this example:

9. Simplify: $\dfrac{x-y}{y-x}$

The expression can be simplified by recognizing that $y-x$ and $x-y$ are opposites:

$y-x = -1(x-y)$.

$\dfrac{x-y}{y-x} = \dfrac{1\cdot(x-y)}{-1\cdot(x-y)} = \dfrac{1}{-1} = \boxed{-1}$

Your turn:

10. Simplify: $\dfrac{x-7}{7-x}$

	Answer	Text Ref	Video Ref		Answer	Text Ref	Video Ref
1	a. $\dfrac{9}{7}$ b. $-\dfrac{2}{7}$	Ex 1, p. 433		**6**	$\dfrac{x^2}{x-2}$		Sec 7.1, 7/9
2	$-\dfrac{1}{4}$		Sec 7.1, 1/9	**7**	$\dfrac{1}{x-9}$	Ex 6, p. 437	
3	$\dfrac{2}{3},1$	Ex 2b, p. 434		**8**	$-(2+x)$ or $-2-x$		Sec 7.1, 8/9
4	$7,-2$		Sec 7.1, 3/9	**9**	-1	Ex 7b, p. 437	
5	$\dfrac{x+2}{x}$	Ex 5, p. 436		**10**	-1		Sec 7.1, 6/9

☐ **Next, insert your homework.** Make sure you attempt all exercises asked of you and show all work, as in the exercises above. Check your answers if possible. Clearly mark any exercises you were unable to correctly complete so that you may ask questions later. DO NOT ERASE YOUR INCORRECT WORK. THIS IS HOW WE UNDERSTAND AND EXPLAIN TO YOU YOUR ERRORS.

Section 7.2 Multiplying and Dividing Rational Expressions

Before Class:

☐   Read the objectives on page 442.

☐   Read the **Helpful Hint** boxes on pages 444, 445, and 448.

☐   Complete the exercises:

1. To multiply rational expressions, multiply the _____ and then

    multiply the _____ .

2. To divide two rational expressions, multiply the first rational expression by the

    _____ of the second rational expression.

During Class:

☐   **Write your class notes.** Neatly write down **all** examples shown as well as key terms or phrases with definitions. If not applicable or if you were absent, watch the Lecture Series (DVD) for this section and do the same (write down the examples shown as well as key terms or phrases). Insert more paper as needed.

Class Notes/Examples	Your Notes

Answers: **1)** numerators, denominators   **2)** reciprocal

Section 7.2 Multiplying and Dividing Rational Expressions

Class Notes (continued)	**Your Notes**

(Insert additional paper as needed.)

Section 7.2 Multiplying and Dividing Rational Expressions

Practice:

☐   Complete the Vocabulary, Readiness & Video Check on page 449.

☐   Next, complete any incomplete exercises below. Check and correct your work using the answers and references at the end of this section.

Review this example:	**Your turn:**
1. Multiply: $\dfrac{-7x^2}{5y} \cdot \dfrac{3y^5}{14x^2}$	**2.** Multiply: $\dfrac{8x}{2} \cdot \dfrac{x^5}{4x^2}$

$\dfrac{-7x^2}{5y} \cdot \dfrac{3y^5}{14x^2} = \dfrac{-7x^2 \cdot 3y^5}{5y \cdot 14x^2}$ Factor the numerator and denominator; divide out common factors

$\qquad\qquad = \dfrac{-1 \cdot 7 \cdot 3 \cdot x^2 \cdot y \cdot y^4}{5 \cdot 2 \cdot 7 \cdot x^2 \cdot y}$

$\qquad\qquad = \boxed{-\dfrac{3y^4}{10}}$

Review this example:	**Your turn:**
3. Multiply: $\dfrac{3x+3}{5x-5x^2} \cdot \dfrac{2x^2+x-3}{4x^2-9}$	**4.** Multiply: $\dfrac{5x-20}{3x^2+x} \cdot \dfrac{3x^2+13x+4}{x^2-16}$

$\dfrac{3x+3}{5x-5x^2} \cdot \dfrac{2x^2+x-3}{4x^2-9} = \dfrac{3(x+1)}{5x(1-x)} \cdot \dfrac{(2x+3)(x-1)}{(2x-3)(2x+3)}$

$\qquad\qquad = \dfrac{3(x+1)(2x+3)(x-1)}{5x(1-x)(2x-3)(2x+3)}$

$\qquad\qquad = \dfrac{3(x+1)(x-1)}{5x(1-x)(2x-3)}$

$\qquad\qquad = \dfrac{3(x+1)(-1)(1-x)}{5x(1-x)(2x-3)}$

$\qquad\qquad = \dfrac{-3(x+1)}{5x(2x-3)}$

$\qquad\qquad \text{or} \boxed{-\dfrac{3(x+1)}{5x(2x-3)}}$

Section 7.2 Multiplying and Dividing Rational Expressions

Review this example:

5. Divide: $\dfrac{(x+2)^2}{10} \div \dfrac{2x+4}{5}$

$\dfrac{(x+2)(x+2)}{10} \div \dfrac{2x+4}{5} = \dfrac{(x+2)(x+2)}{10} \cdot \dfrac{5}{2x+4}$

$= \dfrac{(x+2)(x+2) \cdot 5}{5 \cdot 2 \cdot 2 \cdot (x+2)}$

$= \boxed{\dfrac{x+2}{4}}$

Your turn:

6. Divide: $\dfrac{5x-10}{12} \div \dfrac{4x-8}{8}$

Review this example:

7. Divide: $\dfrac{2x^2-11x+5}{5x-25} \div \dfrac{4x-2}{10}$

$\dfrac{2x^2-11x+5}{5x-25} \div \dfrac{4x-2}{10} = \dfrac{2x^2-11x+5}{5x-25} \cdot \dfrac{10}{4x-2}$

$= \dfrac{(2x-1)(x-5) \cdot 2 \cdot 5}{5(x-5) \cdot 2(2x-1)}$

$= \dfrac{1}{1} = \boxed{1}$

Your turn:

8. Divide: $\dfrac{x+2}{7-x} \div \dfrac{x^2-5x+6}{x^2-9x+14}$

	Answer	Text Ref	Video Ref			Answer	Text Ref	Video Ref
1	$-\dfrac{3y^4}{10}$	Ex 1b, p. 443			5	$\dfrac{x+2}{4}$	Ex 5, p. 445	
2	x^4		Sec 7.2, 1/5		6	$\dfrac{5}{6}$		Sec 7.2, 4/5
3	$-\dfrac{3(x+1)}{5x(2x-3)}$	Ex 3, p. 444			7	1	Ex 7, p. 446	
4	$\dfrac{5}{x}$		Sec 7.2, 2/5		8	$-\dfrac{x+2}{x-3}$		Sec 7.2, 3/5

☐ **Next, insert your homework.** Make sure you attempt all exercises asked of you and show all work, as in the exercises above. Check your answers if possible. Clearly mark any exercises you were unable to correctly complete so that you may ask questions later. DO NOT ERASE YOUR INCORRECT WORK. THIS IS HOW WE UNDERSTAND AND EXPLAIN TO YOU YOUR ERRORS.

Section 7.3 Adding and Subtracting Rational Expressions with Common Denominators and Least Common Denominator

Before Class:

☐ Read the objectives on page 451.

☐ Read the **Helpful Hint** boxes on page 452.

☐ Complete the exercises:

1. To add rational expressions, add the _____ and place the sum

   over _____ .

2. The LCD of a list of rational expressions is a polynomial of least degree whose factors

   include _____ .

During Class:

☐ **Write your class notes.** Neatly write down **all** examples shown as well as key terms or phrases with definitions. If not applicable or if you were absent, watch the Lecture Series (DVD) for this section and do the same (write down the examples shown as well as key terms or phrases). Insert more paper as needed.

Class Notes/Examples	Your Notes

Answers: **1)** numerators, the common denominator **2)** all the factors of the denominators in the list

Section 7.3 Adding and Subtracting Rational Expressions with
Common Denominators and Least Common Denominator

Class Notes (continued)	Your Notes

(Insert additional paper as needed.)

Section 7.3 Adding and Subtracting Rational Expressions with Common Denominators and Least Common Denominator

Practice:

☐   Complete the Vocabulary, Readiness & Video Check on page 456.

☐   Next, complete any incomplete exercises below. Check and correct your work using the answers and references at the end of this section.

Review this example:

1. Subtract: $\dfrac{2y}{2y-7} - \dfrac{7}{2y-7}$

$\dfrac{2y}{2y-7} - \dfrac{7}{2y-7} = \dfrac{2y-7}{2y-7} = \dfrac{1}{1} = \boxed{1}$

Your turn:

2. Add: $\dfrac{9}{3+y} + \dfrac{y+1}{3+y}$

Review this example:

3. Subtract: $\dfrac{3x^2+2x}{x-1} - \dfrac{10x-5}{x-1}$

$\dfrac{3x^2+2x}{x-1} - \dfrac{10x-5}{x-1} = \dfrac{\left(3x^2+2x\right)-\left(10x-5\right)}{x-1}$

$= \dfrac{3x^2+2x-10x+5}{x-1}$

$= \dfrac{3x^2-8x+5}{x-1}$

$= \dfrac{(x-1)(3x-5)}{x-1}$

$= \boxed{3x-5}$

Your turn:

4. Subtract: $\dfrac{2x+3}{x^2-x-30} - \dfrac{x-2}{x^2-x-30}$

Review this example:

5. Find the LCD of $\dfrac{t-10}{t^2-t-6}$ and $\dfrac{t+5}{t^2+3t+2}$.

Start by factoring each denominator.
$t^2-t-6=(t-3)(t+2)$
$t^2+3t+2=(t+1)(t+2)$
$\text{LCD}=\boxed{(t-3)(t+2)(t+1)}$

Your turn:

6. Find the LCD of $\dfrac{1}{3x+3}$ and $\dfrac{8}{2x^2+4x+2}$.

Section 7.3 Adding and Subtracting Rational Expressions with Common Denominators and Least Common Denominator

Review this example:

7. Write the rational expression as an equivalent rational expression with the given denominator.

$$\frac{7x}{2x+5} = \frac{}{6x+15}$$

First, factor the denominator on the right.

$$\frac{7x}{2x+5} = \frac{}{3(2x+5)}$$

To obtain the denominator on the right from the denominator on the left, multiply by 1 in the form of $\frac{3}{3}$.

$$\frac{7x}{2x+5} = \frac{7x}{2x+5} \cdot \frac{3}{3} = \frac{7x \cdot 3}{(2x+5) \cdot 3} = \frac{21x}{3(2x+5)}$$

$$\text{or} \left(\frac{21x}{6x+15}\right)$$

Your turn:

8. Rewrite the rational expression as an equivalent rational expression with the given denominator.

$$\frac{9a+2}{5a+10} = \frac{}{5b(a+2)}$$

	Answer	Text Ref	Video Ref		Answer	Text Ref	Video Ref
1	1	Ex 2, p. 452		5	$(t-3)(t+2)(t+1)$	Ex 7, p. 454	
2	$\frac{y+10}{3+y}$		Sec 7.3, 2/7	6	$6(x+1)^2$		Sec 7.3, 5/7
3	$3x-5$	Ex 3, p. 452		7	$\frac{21x}{6x+15}$	Ex 9b, p. 455	
4	$\frac{1}{x-6}$		Sec 7.3, 3/7	8	$\frac{9ab+2b}{5b(a+2)}$		Sec 7.3, 7/7

☐ **Next, insert your homework.** Make sure you attempt all exercises asked of you and show all work, as in the exercises above. Check your answers if possible. Clearly mark any exercises you were unable to correctly complete so that you may ask questions later. DO NOT ERASE YOUR INCORRECT WORK. THIS IS HOW WE UNDERSTAND AND EXPLAIN TO YOU YOUR ERRORS.

Section 7.4 Adding and Subtracting Rational Expressions with Unlike Denominators

Before Class:

☐   Read the objective on page 459.

☐   Complete the exercises:

1.   Read the Adding or Subtracting Rational Expressions with Unlike Denominators box on page 459. After the LCD is found, what is the second step in this process?

2.   What is the last step in adding or subtracting rational expressions with unlike denominators?

During Class:

☐   **Write your class notes.** Neatly write down **all** examples shown as well as key terms or phrases with definitions. If not applicable or if you were absent, watch the Lecture Series (DVD) for this section and do the same (write down the examples shown as well as key terms or phrases). Insert more paper as needed.

Class Notes/Examples	**Your Notes**

Answers: **1)** Rewrite each rational expression as an equivalent expression whose denominator is the LCD.   **2)** Simplify or write the rational expression in simplest form.

Section 7.4 Adding and Subtracting Rational Expressions with Unlike Denominators

Class Notes (continued)

Your Notes

(Insert additional paper as needed.)

Section 7.4 Adding and Subtracting Rational Expressions with Unlike Denominators

Practice:

☐   Complete the Vocabulary, Readiness & Video Check on page 462.

☐   Next, complete any incomplete exercises below. Check and correct your work using the answers and references at the end of this section.

Review this example:

1. Perform the indicated operation.

$$\frac{3}{10x^2}+\frac{7}{25x}$$

Since $10x^2 = 2 \cdot 5 \cdot x \cdot x$ and $25x = 5 \cdot 5 \cdot x$, the LCD is $2 \cdot 5^2 \cdot x^2 = 50x^2$. Write each fraction as an equivalent fraction with a denominator of $50x^2$.

$$\frac{3}{10x^2}+\frac{7}{25x}=\frac{3(5)}{10x^2(5)}+\frac{7(2x)}{25x(2x)}$$

$$=\frac{15}{50x^2}+\frac{14x}{50x^2}=\boxed{\frac{15+14x}{50x^2}}$$

Your turn:

2. Perform the indicated operation. Simplify if possible.

$$\frac{3}{x}+\frac{5}{2x^2}$$

Review this example:

3. Subtract: $\dfrac{7}{x-3}-\dfrac{9}{3-x}$

To find a common denominator, notice that $x-3$ and $3-x$ are opposites: $3-x=-(x-3)$.

$$\frac{7}{x-3}-\frac{9}{3-x}=\frac{7}{x-3}-\frac{9}{-(x-3)}$$

$$=\frac{7}{x-3}-\frac{-9}{x-3}$$

$$=\frac{7-(-9)}{x-3}=\boxed{\frac{16}{x-3}}$$

Your turn:

4. Perform the indicated operation. Simplify if possible.

$$\frac{6}{x-3}+\frac{8}{3-x}$$

Review this example:

5. Add: $1+\dfrac{m}{m+1}$

The LCD of 1 $\left(1=\dfrac{1}{1}\right)$ and $\dfrac{m}{m+1}$ is $m+1$.

(solution continued on the next page)

Your turn:

6. Perform the indicated operation. Simplify if possible.

$$\frac{y+2}{y+3}-2$$

Section 7.4 Adding and Subtracting Rational Expressions with Unlike Denominators

$$1 + \frac{m}{m+1} = \frac{1}{1} + \frac{m}{m+1}$$

$$= \frac{1(m+1)}{1(m+1)} + \frac{m}{m+1}$$

$$= \frac{m+1+m}{m+1}$$

$$= \boxed{\frac{2m+1}{m+1}}$$

Review this example:

7. Subtract: $\dfrac{3}{2x^2 + x} - \dfrac{2x}{6x+3}$

Factor the denominators.

$$\frac{3}{2x^2+x} - \frac{2x}{6x+3} = \frac{3}{x(2x+1)} - \frac{2x}{3(2x+1)}$$

The LCD is $3x(2x+1)$. Write equivalent expressions with denominators of $3x(2x+1)$.

$$\frac{3(3)}{x(2x+1)(3)} - \frac{2x(x)}{3(2x+1)(x)} = \boxed{\frac{9-2x^2}{3x(2x+1)}}$$

Your turn:

8. Perform the indicated operation. Simplify if possible.

$$\frac{3a}{2a+6} - \frac{a-1}{a+3}$$

	Answer	Text Ref	Video Ref		Answer	Text Ref	Video Ref
1	$\dfrac{15+14x}{50x^2}$	Ex 1b, p. 459		5	$\dfrac{2m+1}{m+1}$	Ex 5, p. 461	
2	$\dfrac{6x+5}{2x^2}$		Sec 7.4, 1/5	6	$-\dfrac{y+4}{y+3}$		Sec 7.4, 3/5
3	$\dfrac{16}{x-3}$	Ex 4, pp. 460–461		7	$\dfrac{9-2x^2}{3x(2x+1)}$	Ex 6, p. 461	
4	$-\dfrac{2}{x-3}$		Sec 7.4, 2/5	8	$\dfrac{a+2}{2(a+3)}$		Sec 7.4, 4/5

☐   **Next, insert your homework.** Make sure you attempt all exercises asked of you and show all work, as in the exercises above. Check your answers if possible. Clearly mark any exercises you were unable to correctly complete so that you may ask questions later. DO NOT ERASE YOUR INCORRECT WORK. THIS IS HOW WE UNDERSTAND AND EXPLAIN TO YOU YOUR ERRORS.

Section 7.5 Solving Equations Containing Rational Expressions

Before Class:

☐   Read the objectives on page 465.

☐   Read the **Helpful Hint** boxes on pages 465, 466, and 468.

☐   Complete the exercises:

     1.   To solve equations containing rational expressions, use the

         _____ of equality to clear the equation of fractions by

         _____ both sides of the equation by the LCD.

     2.   What is the last step to solving an equation containing rational expressions?

During Class:

☐   **Write your class notes.** Neatly write down **all** examples shown as well as key terms or phrases with definitions. If not applicable or if you were absent, watch the Lecture Series (DVD) for this section and do the same (write down the examples shown as well as key terms or phrases). Insert more paper as needed.

Class Notes/Examples	Your Notes

Answers: **1)** multiplication property, multiplying   **2)** Check the solution in the original equation.

Section 7.5 Solving Equations Containing Rational Expressions

Class Notes (continued)

Your Notes

(Insert additional paper as needed.)

Section 7.5 Solving Equations Containing Rational Expressions

Practice:

☐   Complete the Vocabulary, Readiness & Video Check on page 470.

☐   Next, complete any incomplete exercises below. Check and correct your work using the answers and references at the end of this section.

Review this example:

1. Solve: $\dfrac{t-4}{2} - \dfrac{t-3}{9} = \dfrac{5}{18}$

The LCD of denominators 2, 9, and 18 is 18, so we multiply both sides of the equation by 18.

$$18\left(\dfrac{t-4}{2} - \dfrac{t-3}{9}\right) = 18\left(\dfrac{5}{18}\right)$$

$$18\left(\dfrac{t-4}{2}\right) - 18\left(\dfrac{t-3}{9}\right) = 18\left(\dfrac{5}{18}\right)$$

$$9(t-4) - 2(t-3) = 5$$

$$9t - 36 - 2t + 6 = 5$$

$$7t - 30 = 5$$

$$7t = 35$$

$$t = 5$$

Check: $\dfrac{t-4}{2} - \dfrac{t-3}{9} = \dfrac{5}{18}$

$$\dfrac{5-4}{2} - \dfrac{5-3}{9} \overset{?}{=} \dfrac{5}{18}$$

$$\dfrac{1}{2} - \dfrac{2}{9} \overset{?}{=} \dfrac{5}{18}$$

$$\text{True} \quad \dfrac{5}{18} = \dfrac{5}{18}$$

The solution is $\boxed{5.}$

Your turn:

2. Solve: $\dfrac{x-3}{5} + \dfrac{x-2}{2} = \dfrac{1}{2}$

Review this example:

3. Solve: $3 - \dfrac{6}{x} = x + 8$

In this equation, 0 cannnot be a solution because if x is 0, the rational expression $\dfrac{6}{x}$ is undefined. The LCD is x, so multiply both sides of the equation by x.

(solution continued on the next page)

Your turn:

4. Solve: $\dfrac{2}{y} + \dfrac{1}{2} = \dfrac{5}{2y}$

Section 7.5 Solving Equations Containing Rational Expressions

$$x\left(3-\frac{6}{x}\right)=x(x+8)$$

$$x(3)-x\left(\frac{6}{x}\right)=x\cdot x+x\cdot 8$$

$$3x-6=x^2+8x$$

Write the quadratic equation in standard form and solve for x.

$$0=x^2+5x+6$$
$$0=(x+3)(x+2)$$
$$x+3=0 \qquad \text{or} \qquad x+2=0$$
$$x=-3 \qquad\qquad\qquad x=-2$$

Notice that neither -3 nor -2 makes the denominator in the original equation equal to 0.

Check: To check these solutions, replace x in the original equation by -3, and then by -2.

If $x=-3$:

$$3-\frac{6}{x}=x+8$$

$$3-\frac{6}{-3}\overset{?}{=}-3+8$$

$$3-(-2)\overset{?}{=}5$$

$$5=5 \qquad \text{True}$$

If $x=-2$:

$$3-\frac{6}{x}=x+8$$

$$3-\frac{6}{-2}\overset{?}{=}-2+8$$

$$3-(-3)\overset{?}{=}6$$

$$6=6 \qquad \text{True}$$

Both -3 and -2 are solutions.

Review this example:

5. Solve: $\dfrac{4x}{x^2+x-30}+\dfrac{2}{x-5}=\dfrac{1}{x+6}$

The denominator x^2+x-30 factors as $(x+6)(x-5)$. The LCD is then $(x+6)(x-5)$.

$$(x+6)(x-5)\left(\frac{4x}{x^2+x-30}+\frac{2}{x-5}\right)$$

$$=(x+6)(x-5)\left(\frac{1}{x+6}\right)$$

(solution continued on the next page)

Your turn:

6. Solve: $\dfrac{4r-4}{r^2+5r-14}+\dfrac{2}{r+7}=\dfrac{1}{r-2}$

Section 7.5 Solving Equations Containing Rational Expressions

$$(x+6)(x-5)\cdot\frac{4x}{x^2+x-30}+(x+6)(x-5)\cdot\frac{2}{x-5}$$

$$=(x+6)(x-5)\cdot\frac{1}{x+6}$$

$$4x+2(x+6)=x-5$$

$$4x+2x+12=x-5$$

$$6x+12=x-5$$

$$5x=-17$$

$$x=-\frac{17}{5}$$

Check by replacing x with $-\dfrac{17}{5}$ in the original

equation.

The solution is $\boxed{-\dfrac{17}{5}}$.

Review this example:	**Your turn:**

7. Solve: $x+\dfrac{14}{x-2}=\dfrac{7x}{x-2}+1$

8. Solve: $\dfrac{t}{t-4}=\dfrac{t+4}{6}$

Notice the denominators in this equation; 2 can't be a solution. The LCD is $x-2$, so we multiply both sides of the equation by $x-2$.

$$(x-2)\left(x+\frac{14}{x-2}\right)=(x-2)\left(\frac{7x}{x-2}+1\right)$$

$$(x-2)(x)+(x-2)\left(\frac{14}{x-2}\right)$$

$$=(x-2)\left(\frac{7x}{x-2}\right)+(x-2)(1)$$

$$x^2-2x+14=7x+x-2$$

$$x^2-2x+14=8x-2$$

$$x^2-10x+16=0$$

$$(x-8)(x-2)=0$$

$$x-8=0 \text{ or } x-2=0$$

$$x=8 \qquad x=2$$

2 can't be a solution of the original equation. Replace x only with 8 in the original equation.

The only solution is $\boxed{8.}$

Section 7.5 Solving Equations Containing Rational Expressions

Review this example:

9. Solve $\dfrac{1}{a}+\dfrac{1}{b}=\dfrac{1}{x}$ for x.

The LCD is abx, so we multiply both sides by abx.

$$abx\left(\frac{1}{a}+\frac{1}{b}\right)=abx\left(\frac{1}{x}\right)$$

$$abx\left(\frac{1}{a}\right)+abx\left(\frac{1}{b}\right)=abx\cdot\left(\frac{1}{x}\right)$$

$$bx+ax=ab$$

$$x(b+a)=ab$$

$$\frac{x(b+a)}{b+a}=\frac{ab}{b+a}$$

$$\boxed{x=\frac{ab}{b+a}}$$

Your turn:

10. Solve $T=\dfrac{2U}{B+E}$ for B.

	Answer	Text Ref	Video Ref		Answer	Text Ref	Video Ref
1	5	Ex 2, p. 466		6	3		Sec 7.5, 5/6
2	3		Sec 7.5, 1/6	7	8	Ex 6, p. 468	
3	$-3,-2$	Ex 3, pp. 466–467		8	$8,-2$		Sec 7.5, 3/6
4	1		Sec 7.5, 2/6	9	$x=\dfrac{ab}{b+a}$	Ex 7, p. 469	
5	$-\dfrac{17}{5}$	Ex 4, p. 467		10	$B=\dfrac{2U-ET}{T}$		Sec 7.5, 6/6

☐   **Next, insert your homework.** Make sure you attempt all exercises asked of you and show all work, as in the exercises above. Check your answers if possible. Clearly mark any exercises you were unable to correctly complete so that you may ask questions later. DO NOT ERASE YOUR INCORRECT WORK. THIS IS HOW WE UNDERSTAND AND EXPLAIN TO YOU YOUR ERRORS.

Section 7.6 Proportion and Problem Solving with Rational Equations

Before Class:

☐   Read the objectives on page 473.

☐   Read the **Helpful Hint** boxes on pages 476 and 480.

☐   Complete the exercises:

1.   A _____ is the quotient of two numbers or two quantities.

2.   A _____ is a mathematical statement that two ratios are equal.

3.   In similar triangles, corresponding sides are in _____ .

During Class:

☐   **Write your class notes.** Neatly write down **all** examples shown as well as key terms or phrases with definitions. If not applicable or if you were absent, watch the Lecture Series (DVD) for this section and do the same (write down the examples shown as well as key terms or phrases). Insert more paper as needed.

Class Notes/Examples	Your Notes

Answers: **1)** ratio   **2)** proportion   **3)** proportion

Section 7.6 Proportion and Problem Solving with Rational Equations

Class Notes (continued)   **Your Notes**

(Insert additional paper as needed.)

Section 7.6 Proportion and Problem Solving with Rational Equations

Practice:

☐   Complete the Vocabulary, Readiness & Video Check on page 481–482.

☐   Next, complete any incomplete exercises below. Check and correct your work using the answers and references at the end of this section.

Review this example:

1. Solve for x: $\dfrac{x-5}{3} = \dfrac{x+2}{5}$

$$\frac{x-5}{3} = \frac{x+2}{5}$$

$$5(x-5) = 3(x+2)$$

$$5x - 25 = 3x + 6$$

$$5x = 3x + 31$$

$$2x = 31$$

$$\frac{2x}{2} = \frac{31}{2}$$

$$x = \frac{31}{2}$$

Check: Verify that $\left(\dfrac{31}{2}\right)$ is the solution.

Your turn:

2. Solve the proportion.

$$\frac{x+1}{2x+3} = \frac{2}{3}$$

Review this example:

3. Three boxes of CD-Rs (recordable compact discs) cost $37.47. How much should 5 boxes cost?

UNDERSTAND. Read and reread the problem. Let x = price of 5 boxes of CD-Rs.

TRANSLATE.

$$\frac{3 \text{ boxes}}{5 \text{ boxes}} = \frac{\text{price of 3 boxes}}{\text{price of 5 boxes}}$$

$$\frac{3}{5} = \frac{37.47}{x}$$

SOLVE.

$$\frac{3}{5} = \frac{37.47}{x}$$

$$3x = 5(37.47)$$

$$x = 62.45$$

(solution continued on the next page)

Your turn:

4. There are 110 calories per 28.8 grams of Frosted Flakes cereal. Find how many calories are in 43.2 grams of this cereal.

Section 7.6 Proportion and Problem Solving with Rational Equations

INTERPRET.

Check: Verify that 3 boxes are to 5 boxes as $37.47 is to $62.45. Also, notice that our solution is reasonable.

State: Five boxes of CD-Rs cost $62.45.

Review this example:	**Your turn:**

Review this example:

5. The quotient of a number and 6, minus $\frac{5}{3}$, is the quotient of the number and 2. Find the number.

UNDERSTAND. Read and reread the problem. Let x = the unknown number.

TRANSLATE. $\dfrac{x}{6} - \dfrac{5}{3} = \dfrac{x}{2}$

SOLVE. Multiply both sides of the equation by the LCD, 6.

$$6\left(\frac{x}{6} - \frac{5}{3}\right) = 6\left(\frac{x}{2}\right)$$

$$6\left(\frac{x}{6}\right) - 6\left(\frac{5}{3}\right) = 6\left(\frac{x}{2}\right)$$

$$x - 10 = 3x$$

$$-10 = 2x$$

$$\frac{-10}{2} = \frac{2x}{2}$$

$$-5 = x$$

INTERPRET.
Check: Verify that the quotient of -5 and 6 minus $\frac{5}{3}$ is the quotient of -5 and 2, or $-\dfrac{5}{6} - \dfrac{5}{3} = -\dfrac{5}{2}$.

State: The unknown number is -5.

Your turn:

4. Twelve divided by the sum of x and 2 equals the quotient of 4 and the difference of x and 2. Find x.

Section 7.6 Proportion and Problem Solving with Rational Equations

Review this example:

7. Sam and Frank work in a plant that manufactures automobiles. Sam can complete a quality control tour of the plant in 3 hours while his assistant, Frank, needs 7 hours to complete the same job. The regional manager is coming to inspect the plant facilities, so both Sam and Frank are directed to complete a quality control tour together. How long will this take?

UNDERSTAND. Read and reread the problem. The key idea here is the relationship between the *time* (hours) it takes to complete the job and the *part of the job* completed in 1 unit of time (hour).

Let x = the time in hours it takes Sam and Frank to complete the job together.

Then $\dfrac{1}{x}$ = the part of the job they complete in 1 hour.

	Hours to Complete Total Job	*Part of Job Completed in 1 Hour*
Sam	3	$\dfrac{1}{3}$
Frank	7	$\dfrac{1}{7}$
Together	x	$\dfrac{1}{x}$

TRANSLATE.

part of job Sam completed in 1 hour	added to	part of job Frank completed in 1 hour	is equal to	part of job completed together in 1 hour
↓	↓	↓	↓	↓
$\dfrac{1}{3}$	$+$	$\dfrac{1}{7}$	$=$	$\dfrac{1}{x}$

(solution continued on the next page)

Your turn:

8. In 2 minutes, a conveyor belt moves 300 pounds of recyclable aluminum from the delivery truck to a storage area. A smaller belt moves the same quantity of cans the same distance in 6 minutes. If both belts are used, find how long it takes to move the cans to the storage area.

Section 7.6 Proportion and Problem Solving with Rational Equations

SOLVE. Solve the equation $\dfrac{1}{3}+\dfrac{1}{7}=\dfrac{1}{x}$. Multiply both sides of the equation by the LCD, $21x$.

$$21x\left(\frac{1}{3}+\frac{1}{7}\right)=21x\left(\frac{1}{x}\right)$$

$$21x\left(\frac{1}{3}\right)+21x\left(\frac{1}{7}\right)=21x\left(\frac{1}{x}\right)$$

$$7x+3x=21$$

$$10x=21$$

$$x=\frac{21}{10}\ \text{ or }\ 2\frac{1}{10}\ \text{hours}$$

INTERPRET.

Check: Our proposed solution is $2\dfrac{1}{10}$ hours. This

proposed solution is reasonable since $2\dfrac{1}{10}$ hours is

more than half of Sam's time and less than half of Frank's time. Check this solution in the originally stated problem.

State: Sam and Frank can complete the quality

control tour in $\left(2\dfrac{1}{10}\ \text{hours.}\right)$

	Answer	Text Ref	Video Ref		Answer	Text Ref	Video Ref
1	$\dfrac{31}{2}$	Ex 2, p. 475		5	-5	Ex 5, pp. 477–478	
2	-3		Sec 7.6, 2/7	6	4		Sec 7.6, 5/7
3	$62.45	Ex 3, pp. 475–476		7	$2\dfrac{1}{10}$ h	Ex 6, pp. 478–479	
4	165 cal		Sec 7.6, 3/7	8	$1\dfrac{1}{2}$ min		Sec 7.6, 6/7

☐ **Next, insert your homework.** Make sure you attempt all exercises asked of you and show all work, as in the exercises above. Check your answers if possible. Clearly mark any exercises you were unable to correctly complete so that you may ask questions later. DO NOT ERASE YOUR INCORRECT WORK. THIS IS HOW WE UNDERSTAND AND EXPLAIN TO YOU YOUR ERRORS.

Section 7.7 Variation and Problem Solving

Before Class:

☐   Read the objectives on page 486.

☐   Read the **Helpful Hint** box on page 491.

☐   Complete the exercises:

     1. y varies _____ as x if there is a nonzero constant k such that $y = kx$.

     2. y varies _____ as x if there is a nonzero constant k such that $y = \dfrac{k}{x}$.

During Class:

☐   **Write your class notes.** Neatly write down **all** examples shown as well as key terms or phrases with definitions. If not applicable or if you were absent, watch the Lecture Series (DVD) for this section and do the same (write down the examples shown as well as key terms or phrases). Insert more paper as needed.

         Class Notes/Examples   **Your Notes**

Answers: **1)** directly   **2)** inversely

Section 7.7 Variation and Problem Solving

Class Notes (continued)	Your Notes

(Insert additional paper as needed.)

Section 7.7 Variation and Problem Solving

Practice:

☐   Complete the Vocabulary, Readiness & Video Check on page 493–494.

☐   Next, complete any incomplete exercises below. Check and correct your work using the answers and references at the end of this section.

Review this example:

1. Write a direct variation equation of the form $y = kx$ that satisfies the given ordered pairs.

x	2	9	1.5	-1
y	6	27	4.5	-3

There is a direct variation relationship between x and y. This means that $y = kx$. To find k, substitute one given ordered pair into this equation and solve for k. Use $(2, 6)$.

$y = kx$

$6 = k \cdot 2$

$\dfrac{6}{2} = \dfrac{k \cdot 2}{2}$

$3 = k$

Since $k = 3$, $y = 3x$.

Check: See that each given y is 3 times the given x.

Your turn:

2. Write a direct variation equation, $y = kx$, that satisfies the ordered pairs in the table.

x	-2	2	4	5
y	-12	12	24	30

Review this example:

3. Suppose that y varies directly as x. If y is 17 when x is 34, find the constant of variation and the direct variation equation. Then find y when x is 12.

The relationship is of the form $y = kx$. Let $y = 17$ and $x = 34$ and solve for k.

$17 = k \cdot 34$

$\dfrac{17}{34} = \dfrac{k \cdot 34}{34}$

$\dfrac{1}{2} = k$ Thus the equation is $y = \dfrac{1}{2}x$.

To find y when $x = 12$, replace x with 12 in $y = \dfrac{1}{2}x$.

$y = \dfrac{1}{2}x = \dfrac{1}{2} \cdot 12 = 6$ Thus, when x is 12, y is 6.

Your turn:

4. y varies directly as x. If $y = 20$ when $x = 5$, find y when x is 10.

Section 7.7 Variation and Problem Solving

Review this example:

5. Write an inverse variation equation of the form $y = \dfrac{k}{x}$ that satisfies the ordered pairs in the table.

x	2	4	$\dfrac{1}{2}$
y	6	3	24

To find k, choose one given ordered pair and substitute the values into the equation. Use $(2,6)$.

$$y = \frac{k}{x}$$

$$6 = \frac{k}{2}$$

$$2 \cdot 6 = 2 \cdot \frac{k}{2}$$

$$12 = k$$

Since $k = 12$, $\boxed{y = \dfrac{12}{x}}$.

Your turn:

6. Write an inverse variation equation, $y = \dfrac{k}{x}$, that satisfies the ordered pairs in the table.

x	1	−7	3.5	−2
y	7	−1	2	−3.5

Review this example:

7. Suppose that y varies inversely as x. If $y = 0.02$ when $x = 75$, find the constant of variation and the inverse variation equation. Then find y when x is 30.

Since y varies inversely as x, the constant of variation may be found by simply finding the product of the given x and y.

$$k = xy = 75(0.02) = 1.5$$

Thus, the equation is $y = \dfrac{1.5}{x}$. To find y when $x = 30$, replace x with 30 in $y = \dfrac{1.5}{x}$.

$$y = \frac{1.5}{x}$$

$$y = \frac{1.5}{30}$$

$$y = 0.05$$

Thus, when x is 30, y is $\boxed{0.05}$.

Your turn:

8. y varies inversely as x. If $y = 5$ when $x = 60$, find y when x is 100.

Section 7.7 Variation and Problem Solving

Review this example:

9. The weight of a body *w* varies inversely with the square of its distance from the center of Earth, *d*. If a person weighs 160 pounds on the surface of Earth, what is the person's weight 200 miles above the surface? (Assume that the radius of Earth is 4000 miles.)

UNDERSTAND. Read and reread the problem.

TRANSLATE. Since weight, *w*, varies inversely with the square of its distance from the center of Earth, *d*,

$$w = \frac{k}{d^2}.$$

SOLVE. First find *k*. Use the fact that the person weighs 160 pounds on Earth's surface, which is a distance of 4000 miles from Earth's center.

$$w = \frac{k}{d^2}$$

$$160 = \frac{k}{(4000)^2}$$

$$2,560,000,000 = k$$

Thus $w = \frac{2,560,000,000}{d^2}$.

To know the person's weight 200 miles above the Earth's surface, let $d = 4200$ and find *w*.

$$w = \frac{2,560,000,000}{d^2}$$

$$w = \frac{2,560,000,000}{(4200)^2}$$

$$w \approx 145$$

INTERPRET.
Check: Your answer is reasonable since the farther a person is from Earth, the less the person weighs.

State: 200 miles above Earth's surface, a 160-pound person weighs approximately 145 pounds.

Your turn:

10. The distance a spring stretches varies directly with the weight attached to the spring. If a 60-pound weight stretches the spring 4 inches, find the distance that an 80-pound weight stretches the spring.

Section 7.7 Variation and Problem Solving

	Answer	Text Ref	Video Ref		Answer	Text Ref	Video Ref
1	$y = 3x$	Ex 1, p. 487		**6**	$y = \dfrac{7}{x}$		Sec 7.7, 3/8
2	$y = 6x$		Sec 7.7, 1/8	**7**	$y = 0.05$	Ex 5, p. 491	
3	$y = 6$	Ex 2, p. 488		**8**	$y = 3$		Sec 7.7, 4/8
4	$y = 40$		Sec 7.7, 2/8	**9**	≈ 145 lb	Ex 7, pp. 492–493	
5	$y = \dfrac{12}{x}$	Ex 4, p. 490		**10**	$5\dfrac{1}{3}$ in.		Sec 7.7, 8/8

☐   **Next, insert your homework.** Make sure you attempt all exercises asked of you and show all work, as in the exercises above. Check your answers if possible. Clearly mark any exercises you were unable to correctly complete so that you may ask questions later. DO NOT ERASE YOUR INCORRECT WORK. THIS IS HOW WE UNDERSTAND AND EXPLAIN TO YOU YOUR ERRORS.

Section 7.8 Simplifying Complex Fractions

Before Class:

☐   Read the objectives on page 496.

☐   Read the **Helpful Hint** box on page 499.

☐   Complete the exercises:

    1.   Read the Method 1: Simplifying a Complex Fraction box on page 498. After the fractions in the numerator or denominator have been added or subtracted, what must be done next to simplify the complex fraction?

    2.   Read the Method 2: Simplifying a Complex Fraction box on page 496. After the LCD of all the fractions in the complex fraction has been found, what must be done next to simplify the complex fraction?

During Class:

☐   **Write your class notes.** Neatly write down **all** examples shown as well as key terms or phrases with definitions. If not applicable or if you were absent, watch the Lecture Series (DVD) for this section and do the same (write down the examples shown as well as key terms or phrases). Insert more paper as needed.

Class Notes/Examples	Your Notes

Answers: **1)** Perform the indicated division by multiplying the numerator of the complex fraction by the reciprocal of the denominator of the complex fraction.   **2)** Multiply both the numerator and the denominator of the complex fraction by the LCD.

Section 7.8 Simplifying Complex Fractions

Class Notes (continued)	**Your Notes**

(Insert additional paper as needed.)

Practice:

☐   Complete the Vocabulary, Readiness & Video Check on page 500.

☐   Next, complete any incomplete exercises below. Check and correct your work using the answers and references at the end of this section.

Review this example:

1. Simplify: $\dfrac{\dfrac{2}{3}+\dfrac{1}{5}}{\dfrac{2}{3}-\dfrac{2}{9}}$

     (Use Method 1 for Simplifying a Complex Fraction)

First add $\dfrac{2}{3}$ and $\dfrac{1}{5}$ to obtain a single fraction in the

numerator; then subtract $\dfrac{2}{9}$ from $\dfrac{2}{3}$ to obtain a

single fraction in the denominator.

$\dfrac{\dfrac{2}{3}+\dfrac{1}{5}}{\dfrac{2}{3}-\dfrac{2}{9}}=\dfrac{\dfrac{2(5)}{3(5)}+\dfrac{1(3)}{5(3)}}{\dfrac{2(3)}{3(3)}-\dfrac{2}{9}}$

$=\dfrac{\dfrac{10}{15}+\dfrac{3}{15}}{\dfrac{6}{9}-\dfrac{2}{9}}$

$=\dfrac{\dfrac{13}{15}}{\dfrac{4}{9}}$

Perform the indicated division by multiplying the numerator of the complex fraction by the reciprocal of the denominator.

$\dfrac{\dfrac{13}{15}}{\dfrac{4}{9}}=\dfrac{13}{15}\cdot\dfrac{9}{4}=\dfrac{13\cdot3\cdot3}{3\cdot5\cdot4}=\boxed{\dfrac{39}{20}}$

Your turn:

2. Simplify the complex fraction using method 1.

     $\dfrac{\dfrac{1}{5}-\dfrac{1}{x}}{\dfrac{7}{10}+\dfrac{1}{x^2}}$

Section 7.8 Simplifying Complex Fractions

Review this example:

3. Simplify: $\dfrac{\dfrac{2}{3}+\dfrac{1}{5}}{\dfrac{2}{3}-\dfrac{2}{9}}$

(Use Method 2 for Simplifying a Complex Fraction)

The LCD of $\dfrac{2}{3}, \dfrac{1}{5}, \dfrac{2}{3}, \dfrac{2}{9}$ is 45, so multiply the numerator and the denominator of the complex fraction by 45. Then perform the indicated operations, and write in the simplest form.

$$\frac{\dfrac{2}{3}+\dfrac{1}{5}}{\dfrac{2}{3}-\dfrac{2}{9}} = \frac{45\left(\dfrac{2}{3}+\dfrac{1}{5}\right)}{45\left(\dfrac{2}{3}-\dfrac{2}{9}\right)}$$

$$= \frac{45\left(\dfrac{2}{3}\right)+45\left(\dfrac{1}{5}\right)}{45\left(\dfrac{2}{3}\right)-45\left(\dfrac{2}{9}\right)}$$

$$= \frac{30+9}{30-10} = \boxed{\frac{39}{20}}$$

Your turn:

4. Simplify the complex fraction using method 2.

$$\frac{\dfrac{1}{5}-\dfrac{1}{x}}{\dfrac{7}{10}+\dfrac{1}{x^2}}$$

	Answer	Text Ref	Video Ref			Answer	Text Ref	Video Ref
1	$\dfrac{39}{20}$	Ex 2, p. 497			**3**	$\dfrac{39}{20}$	Ex 4, pp. 498–499	
2	$\dfrac{2x(x-5)}{7x^2+10}$		Sec 7.8, 3/5		**4**	$\dfrac{2x(x-5)}{7x^2+10}$		Sec 7.8, 4/5

☐ **Next, insert your homework.** Make sure you attempt all exercises asked of you and show all work, as in the exercises above. Check your answers if possible. Clearly mark any exercises you were unable to correctly complete so that you may ask questions later. DO NOT ERASE YOUR INCORRECT WORK. THIS IS HOW WE UNDERSTAND AND EXPLAIN TO YOU YOUR ERRORS.

Preparing for the Chapter 7 Test

Start preparing for your Chapter 7 Test as soon as possible. Pay careful attention to any instructor discussion about this test, especially discussion on what sections you will be responsible for, etc.

☐   Work the Chapter 7 Vocabulary Check on page 502.

☐   Read your Class Notes/Examples for each section covered on your Chapter 7 Test. Look for any unresolved questions you may have.

☐   Complete as many of the Chapter 7 Review exercises as possible (page 507). Remember, the odd answers are in the back of your text.

☐   **Most important:** Place yourself in "test" conditions (see below) and work the Chapter 7 Test (page 509) as a practice test the day before your actual test. To honestly assess how you are doing, try the following:
- Work on a few blank sheets of paper.
- Give yourself the same amount of time you will be given for your actual test.
- Complete this Chapter 7 Practice Test without using your notes or your text.
- If you have any time left after completing this practice test, check your work and try to find any errors on your own.
- Once done, use the back of your book to check ALL answers.
- Try to correct any errors on your own.
- Use the Chapter Test Prep Video (CTPV) to correct any errors you were unable to correct on your own. You can find these videos in the Interactive DVD Lecture Series, in MyMathLab, and on YouTube. Search Martin-Gay Beginning Algebra and click "Channels."

I wish you the best of luck....Elayn Martin-Gay

Section 8.1 Introduction to Radicals

Before Class:

☐   Read the objectives on page 513.

☐   Read the **Helpful Hint** box on page 515.

☐   Complete the exercises:

1.   A number b is a square root of a number a if _____ .

2.   A square root of a _____ number is not a real number.

3.   For any real number a, $\sqrt{a^2} =$ _____ .

During Class:

☐   **Write your class notes.** Neatly write down **all** examples shown as well as key terms or phrases with definitions. If not applicable or if you were absent, watch the Lecture Series (DVD) for this section and do the same (write down the examples shown as well as key terms or phrases). Insert more paper as needed.

Class Notes/Examples	**Your Notes**

Answers: **1)** $b^2 = a$ **2)** negative   **3)** $|a|$

Section 8.1 Introduction to Radicals

Class Notes (continued)

Your Notes

(Insert additional paper as needed.)

Practice:

☐ Complete the Vocabulary, Readiness & Video Check on page 518.

☐ Next, complete any incomplete exercises below. Check and correct your work using the answers and references at the end of this section.

Review this example: **1.** Find each square root. a. $\sqrt{36}$ b. $-\sqrt{16}$ a. $\sqrt{36} = \boxed{6},$ because $6^2 = 36$ and 6 is positive. b. $-\sqrt{16} = \boxed{-4.}$ The negative sign in front of the radical indicates the negative square root of 16.	**Your turn:** **2.** Find each square root. a. $\sqrt{49}$ b. $\sqrt{-4}$
Review this example: **3.** Find each cube root. a. $\sqrt[3]{1}$ b. $\sqrt[3]{\dfrac{1}{125}}$ a. $\sqrt[3]{1} = \boxed{1}$ because $1^3 = 1$. b. $\sqrt[3]{\dfrac{1}{125}} = \boxed{\dfrac{1}{5}}$ because $\left(\dfrac{1}{5}\right)^3 = \dfrac{1}{125}$.	**Your turn:** **4.** Find each cube root. a. $\sqrt[3]{8}$ b. $\sqrt[3]{125}$
Review this example: **5.** Find each root. a. $\sqrt[4]{16}$ b. $-\sqrt[3]{8}$ a. $\sqrt[4]{16} = \boxed{2}$ because $2^4 = 16$ and 2 is positive. b. $-\sqrt[3]{8} = \boxed{-2}$ since $\sqrt[3]{8} = 2$.	**Your turn:** **6.** Find each root. a. $\sqrt[4]{81}$ b. $-\sqrt[5]{32}$
Review this example: **7.** Use an appendix or a calculator to approximate $\sqrt{3}$ to three decimal places. To use a calculator, find the square root key $\boxed{\sqrt{}}$. $\sqrt{3} \approx 1.732050808$ To three decimal places, $\sqrt{3} \approx \boxed{1.732}$	**Your turn:** **8.** Use a calculator or an appendix to approximate $\sqrt{136}$ to three decimal places.

Section 8.1 Introduction to Radicals

Review this example:

9. Simplify each expression. Assume that all variables represent positive numbers.

   a. $\sqrt{x^6}$ b. $\sqrt{\dfrac{x^4}{25}}$ c. $\sqrt[3]{-125a^{12}b^{15}}$

   a. $\sqrt{x^6} = \left(x^3\right)$ because $\left(x^3\right)^2 = x^6$.

   b. $\sqrt{\dfrac{x^4}{25}} = \left(\dfrac{x^2}{5}\right)$ because $\left(\dfrac{x^2}{5}\right)^2 = \dfrac{x^4}{25}$.

   c. $\sqrt[3]{-125a^{12}b^{15}} = \left(-5a^4b^5\right)$ because $\left(-5a^4b^5\right)^3 = -125a^{12}b^{15}$.

Your turn:

10. Simplify each expression. Assume that all variables represent positive numbers.

   a. $\sqrt{x^4}$

   b. $\sqrt{\dfrac{x^6}{36}}$

   c. $\sqrt[3]{a^6b^{18}}$

	Answer	Text Ref	Video Ref		Answer	Text Ref	Video Ref
1	a. 6 b. −4	Ex 1a, b, p. 513		6	a. 3 b. −2		Sec 8.1, 12/21, 14/21
2	a. 7 b. not a real number		Sec 8.1, 1–2/21, 8/21	7	1.732	Ex 4 p. 516	
3	a. 1 b. $\dfrac{1}{5}$	Ex 2a, c, pp. 514–515		8	11.662		Sec 8.1, 16/21
4	a. 2 b. 5		Sec 8.1, 9–10/21	9	a. x^3 b. $\dfrac{x^2}{5}$ c. $-5a^4b^5$	Ex 5b, e, f, pp. 516–517	
5	a. 2 b. −2	Ex 3a, c, p. 515		10	a. x^2 b. $\dfrac{x^3}{6}$ c. a^2b^6		Sec 8.1, 18/21, 20–21/21

☐   **Next, insert your homework.** Make sure you attempt all exercises asked of you and show all work, as in the exercises above. Check your answers if possible. Clearly mark any exercises you were unable to correctly complete so that you may ask questions later. DO NOT ERASE YOUR INCORRECT WORK. THIS IS HOW WE UNDERSTAND AND EXPLAIN TO YOU YOUR ERRORS.

Section 8.2 Simplifying Radicals

Before Class:

☐   Read the objectives on page 520.

☐   Read the **Helpful Hint** box on page 521.

☐   Complete the exercises:

    1.   The simplest form of a radical expression is a(n) _____ form and does not mean a decimal approximation.

    2.   The square root of a product is equal to what?

    3.   The square root of a quotient is equal to what?

During Class:

☐   **Write your class notes.** Neatly write down **all** examples shown as well as key terms or phrases with definitions. If not applicable or if you were absent, watch the Lecture Series (DVD) for this section and do the same (write down the examples shown as well as key terms or phrases). Insert more paper as needed.

Class Notes/Examples	Your Notes

Answers: **1)** exact   **2)** the product of the square roots   **3)** the quotient of the square roots

Section 8.2 Simplifying Radicals

Class Notes (continued)	Your Notes

(Insert additional paper as needed.)

Practice:

☐   Complete the Vocabulary, Readiness & Video Check on page 524–525.

☐   Next, complete any incomplete exercises below. Check and correct your work using the answers and references at the end of this section.

Review this example:

1. Simplify: $\sqrt{12}$

$\sqrt{12} = \sqrt{4 \cdot 3}$
$= \sqrt{4} \cdot \sqrt{3}$
$= \boxed{2\sqrt{3}}$

Your turn:

2. Use the product rule to simplify $\sqrt{20}$.

Review this example:

3. Simplify: $3\sqrt{8}$

Remember that $3\sqrt{8}$ means $3 \cdot \sqrt{8}$.

$3 \cdot \sqrt{8} = 3 \cdot \sqrt{4 \cdot 2}$
$= 3 \cdot \sqrt{4} \cdot \sqrt{2}$
$= 3 \cdot 2 \cdot \sqrt{2} = \boxed{6\sqrt{2}}$

Your turn:

4. Use the product rule to simplify $-5\sqrt{27}$.

Review this example:

5. Simplify: $\sqrt{\dfrac{40}{81}}$

$\sqrt{\dfrac{40}{81}} = \dfrac{\sqrt{40}}{\sqrt{81}}$

$= \dfrac{\sqrt{4} \cdot \sqrt{10}}{9} = \boxed{\dfrac{2\sqrt{10}}{9}}$

Your turn:

6. Use the quotient rule to simplify

$\sqrt{\dfrac{27}{121}}$.

Review this example:

7. Simplify. Assume that all variables represent positive numbers.

a. $\sqrt{x^5}$ b. $\sqrt{\dfrac{45}{x^6}}$

a. $\sqrt{x^5} = \sqrt{x^4 \cdot x} = \sqrt{x^4} \cdot \sqrt{x} = \boxed{x^2 \sqrt{x}}$

b. $\sqrt{\dfrac{45}{x^6}} = \dfrac{\sqrt{45}}{\sqrt{x^6}} = \dfrac{\sqrt{9 \cdot 5}}{x^3} = \dfrac{\sqrt{9} \cdot \sqrt{5}}{x^3} = \boxed{\dfrac{3\sqrt{5}}{x^3}}$

Your turn:

8. Simplify each radical. Assume that all variables represent positive numbers.

a. $\sqrt{x^{13}}$

b. $\sqrt{\dfrac{12}{m^2}}$

Section 8.2 Simplifying Radicals

Review this example:
9. Simplify.

   a. $\sqrt[3]{18}$ b. $\sqrt[4]{\dfrac{3}{16}}$

a. The number 18 contains no perfect cube factors,

   so $\boxed{\sqrt[3]{18}}$ cannot be simplified further.

b. $\sqrt[4]{\dfrac{3}{16}} = \dfrac{\sqrt[4]{3}}{\sqrt[4]{16}} = \boxed{\dfrac{\sqrt[4]{3}}{2}}$

Your turn:
10. Simplify each radical.

   a. $\sqrt[3]{250}$

   b. $\sqrt[4]{\dfrac{8}{81}}$

	Answer	Text Ref	Video Ref		Answer	Text Ref	Video Ref
1	$2\sqrt{3}$	Ex 1b, p. 521		6	$\dfrac{3\sqrt{3}}{11}$		Sec 8.2, 5/11
2	$2\sqrt{5}$		Sec 8.2, 1/11	7	a. $x^2\sqrt{x}$ b. $\dfrac{3\sqrt{5}}{x^3}$	Ex 4a, c, p. 523	
3	$6\sqrt{2}$	Ex 2, p. 522		8	a. $x^6\sqrt{x}$ b. $\dfrac{2\sqrt{3}}{m}$		Sec 8.2, 6/11, 8/11
4	$-15\sqrt{3}$		Sec 8.2, 4/11	9	a. $\sqrt[3]{18}$ b. $\dfrac{\sqrt[4]{3}}{2}$	Ex 5b, 6b, p. 524	
5	$\dfrac{2\sqrt{10}}{9}$	Ex 3c, p. 522		10	a. $5\sqrt[3]{2}$ b. $\dfrac{\sqrt[4]{8}}{3}$		Sec 8.2, 9/11, 11/11

☐   **Next, insert your homework.** Make sure you attempt all exercises asked of you and show all work, as in the exercises above. Check your answers if possible. Clearly mark any exercises you were unable to correctly complete so that you may ask questions later. DO NOT ERASE YOUR INCORRECT WORK. THIS IS HOW WE UNDERSTAND AND EXPLAIN TO YOU YOUR ERRORS.

Section 8.3 Adding and Subtracting Radicals

Before Class:

☐   Read the objectives on page 527.

☐   Complete the exercises:

1. To combine like radicals, use the _____ property.

2. Like radicals have the same _____ and the same

    _____ .

During Class:

☐   **Write your class notes.** Neatly write down **all** examples shown as well as key terms or phrases with definitions. If not applicable or if you were absent, watch the Lecture Series (DVD) for this section and do the same (write down the examples shown as well as key terms or phrases). Insert more paper as needed.

Class Notes/Examples	Your Notes

Answers: **1)** distributive   **2)** index, radicand

Section 8.3 Adding and Subtracting Radicals

Class Notes (continued)

Your Notes

(Insert additional paper as needed.)

Section 8.3 Adding and Subtracting Radicals

Practice:

☐   Complete the Vocabulary, Readiness & Video Check on page 529–530.

☐   Next, complete any incomplete exercises below. Check and correct your work using the
       answers and references at the end of this section.

Review this example:

1. Simplify by combining like radical terms.

   a. $\sqrt{10} - 6\sqrt{10}$ b. $\sqrt[3]{7} + \sqrt[3]{7} - 4\sqrt[3]{5}$

a. $\sqrt{10} - 6\sqrt{10} = 1\sqrt{10} - 6\sqrt{10} = (1-6)\sqrt{10}$

   $= \boxed{-5\sqrt{10}}$

b. $\sqrt[3]{7} + \sqrt[3]{7} - 4\sqrt[3]{5} = 1\sqrt[3]{7} + 1\sqrt[3]{7} - 4\sqrt[3]{5}$

   $= (1+1)\sqrt[3]{7} - 4\sqrt[3]{5}$

   $= \boxed{2\sqrt[3]{7} - 4\sqrt[3]{5}}$

Your turn:

2. Simplify each expression by combining like radicals where possible.

   a. $3\sqrt{6} + 8\sqrt{6} - 2\sqrt{6} - 5$

   b. $2\sqrt[3]{2} - 7\sqrt[3]{2} - 6$

Review this example:

3. Subtract by first simplifying each radical.

   $7\sqrt{12} - \sqrt{75}$

$7\sqrt{12} - \sqrt{75} = 7\sqrt{4 \cdot 3} - \sqrt{25 \cdot 3}$

   $= 7\sqrt{4} \cdot \sqrt{3} - \sqrt{25} \cdot \sqrt{3}$

   $= 7 \cdot 2\sqrt{3} - 5\sqrt{3}$

   $= 14\sqrt{3} - 5\sqrt{3}$

   $= \boxed{9\sqrt{3}}$

Your turn:

4. Add by first simplifying each radical and then combining any like radical terms.

   $\sqrt{12} + \sqrt{27}$

Review this example:

5. Simplify $2\sqrt{x^2} - \sqrt{25x^5} + \sqrt{x^5}$. Assume variables represent positive numbers.

$2\sqrt{x^2} - \sqrt{25x^5} + \sqrt{x^5} = 2x - \sqrt{25x^4 \cdot x} + \sqrt{x^4 \cdot x}$

   $= 2x - \sqrt{25x^4} \cdot \sqrt{x} + \sqrt{x^4} \cdot \sqrt{x}$

   $= 2x - 5x^2\sqrt{x} + x^2\sqrt{x}$

   $= \boxed{2x - 4x^2\sqrt{x}}$

Your turn:

6. Add by first simplifying each radical and then combining any like radical terms. Assume variables represent positive numbers.

   $5\sqrt{2x} + \sqrt{98x}$

Section 8.3 Adding and Subtracting Radicals

Review this example:

7. Simplify the radical expression:

$$5\sqrt[3]{16x^3} - \sqrt[3]{54x^3}$$

$$
\begin{aligned}
5\sqrt[3]{16x^3} - \sqrt[3]{54x^3} &= 5\sqrt[3]{8x^3 \cdot 2} - \sqrt[3]{27x^3 \cdot 2} \\
&= 5 \cdot \sqrt[3]{8x^3} \cdot \sqrt[3]{2} - \sqrt[3]{27x^3} \cdot \sqrt[3]{2} \\
&= 5 \cdot 2x \cdot \sqrt[3]{2} - 3x \cdot \sqrt[3]{2} \\
&= 10x\sqrt[3]{2} - 3x\sqrt[3]{2} \\
&= \boxed{7x\sqrt[3]{2}}
\end{aligned}
$$

Your turn:

8. Add by first simplifying each radical and then combining any like radical terms.

$$\sqrt[3]{8} + \sqrt[3]{54} - 5$$

	Answer	Text Ref	Video Ref		Answer	Text Ref	Video Ref
1	a. $-5\sqrt{10}$ b. $2\sqrt[3]{7} - 4\sqrt[3]{5}$	Ex 1b, c, p. 528		**5**	$2x - 4x^2\sqrt{x}$	Ex 3, p. 529	
2	a. $9\sqrt{6} - 5$ b. $-5\sqrt[3]{2} - 6$		Sec 8.3, 1/8, 4/8	**6**	$12\sqrt{2x}$		Sec 8.3, 6/8
3	$9\sqrt{3}$	Ex 2b, p. 528		**7**	$7x\sqrt[3]{2}$	Ex 4, p. 529	
4	$5\sqrt{3}$		Sec 8.3, 5/8	**8**	$-3 + 3\sqrt[3]{2}$		Sec 8.3, 8/8

☐ **Next, insert your homework.** Make sure you attempt all exercises asked of you and show all work, as in the exercises above. Check your answers if possible. Clearly mark any exercises you were unable to correctly complete so that you may ask questions later. DO NOT ERASE YOUR INCORRECT WORK. THIS IS HOW WE UNDERSTAND AND EXPLAIN TO YOU YOUR ERRORS.

Section 8.4 Multiplying and Dividing Radicals

Before Class:

☐ Read the objectives on page 532.

☐ Read the **Helpful Hint** box on page 537.

☐ Complete the exercises:

1. The product of the *n*th roots of two numbers is the *n*th root of what?

2. To rationalize a denominator that is a sum, multiply the numerator and the denominator

    by the _____ of the sum.

During Class:

☐ **Write your class notes.** Neatly write down **all** examples shown as well as key terms or phrases with definitions. If not applicable or if you were absent, watch the Lecture Series (DVD) for this section and do the same (write down the examples shown as well as key terms or phrases). Insert more paper as needed.

Class Notes/Examples	Your Notes

Answers: **1)** the product of the two numbers   **2)** conjugate

Section 8.4 Multiplying and Dividing Radicals

Class Notes (continued)   **Your Notes**

(Insert additional paper as needed.)

Copyright © 2013 Pearson Education, Inc.

Section 8.4 Multiplying and Dividing Radicals

Practice:

☐ Complete the Vocabulary, Readiness & Video Check on page 538.

☐ Next, complete any incomplete exercises below. Check and correct your work using the answers and references at the end of this section.

Review this example:

1. Find $\left(3\sqrt{2}\right)^2$.

$$\left(3\sqrt{2}\right)^2 = 3^2 \cdot \left(\sqrt{2}\right)^2 = 9 \cdot 2 = \boxed{18}$$

Your turn:

2. Multiply and simplify. Assume that the variable represents a positive real number.

$$\left(6\sqrt{x}\right)^2$$

Review this example:

3. Multiply. Then simplify if possible.

$$\sqrt{5}\left(\sqrt{5} - \sqrt{2}\right)$$

Using the distributive property, we have

$$\sqrt{5}\left(\sqrt{5} - \sqrt{2}\right) = \sqrt{5} \cdot \sqrt{5} - \sqrt{5} \cdot \sqrt{2} = \boxed{5 - \sqrt{10}}$$

Your turn:

4. Multiply and simplify.

$$\sqrt{6}\left(\sqrt{5} + \sqrt{7}\right)$$

Review this example:

5. Divide. Then simplify if possible.

a. $\dfrac{\sqrt{100}}{\sqrt{5}}$ b. $\dfrac{\sqrt{12x^3}}{\sqrt{3x}}$

Use the quotient rule and then simplify the resulting radicand.

a. $\dfrac{\sqrt{100}}{\sqrt{5}} = \sqrt{\dfrac{100}{5}} = \sqrt{20} = \sqrt{4 \cdot 5} = \sqrt{4} \cdot \sqrt{5} = \boxed{2\sqrt{5}}$

b. $\dfrac{\sqrt{12x^3}}{\sqrt{3x}} = \sqrt{\dfrac{12x^3}{3x}} = \sqrt{4x^2} = \boxed{2x}$

Your turn:

6. Divide and simplify. Assume that all variables represent positive real numbers.

a. $\dfrac{\sqrt{90}}{\sqrt{5}}$

b. $\dfrac{\sqrt{75y^5}}{\sqrt{3y}}$

Section 8.4 Multiplying and Dividing Radicals

Review this example:

7. Rationalize the denominator.

$$\sqrt{\frac{1}{18x}}$$

First simplify.

$$\sqrt{\frac{1}{18x}} = \frac{\sqrt{1}}{\sqrt{18x}} = \frac{1}{\sqrt{9} \cdot \sqrt{2x}} = \frac{1}{3\sqrt{2x}}$$

To rationalize the denominator, multiply by $\dfrac{\sqrt{2x}}{\sqrt{2x}}$.

$$\frac{1}{3\sqrt{2x}} = \frac{1}{3\sqrt{2x}} \cdot \frac{\sqrt{2x}}{\sqrt{2x}} = \frac{1 \cdot \sqrt{2x}}{3\sqrt{2x} \cdot \sqrt{2x}} = \frac{\sqrt{2x}}{3 \cdot 2x} = \boxed{\frac{\sqrt{2x}}{6x}}$$

Your turn:

8. Rationalize the denominator and simplify. Assume that all variables represent positive real numbers.

$$\sqrt{\frac{y}{12x}}$$

Review this example:

9. Rationalize the denominator and simplify.

$$\frac{3}{1+\sqrt{x}}$$

$$\frac{3}{1+\sqrt{x}} = \frac{3(1-\sqrt{x})}{(1+\sqrt{x})(1-\sqrt{x})} = \boxed{\frac{3(1-\sqrt{x})}{1-x}}$$

Your turn:

10. Rationalize the denominator and simplify.

$$\frac{4}{2-\sqrt{5}}$$

	Answer	Text Ref	Video Ref		Answer	Text Ref	Video Ref
1	18	Ex 2, p. 533		6	a. $3\sqrt{2}$ b. $5y^2$		Sec 8.4, 6–7/11
2	$36x$		Sec 8.4, 2/11	7	$\dfrac{\sqrt{2x}}{6x}$	Ex 8c, p. 535	
3	$5-\sqrt{10}$	Ex 4a, p. 533		8	$\dfrac{\sqrt{3xy}}{6x}$		Sec 8.4, 9/11
4	$\sqrt{30}+\sqrt{42}$		Sec 8.4, 4/11	9	$\dfrac{3(1-\sqrt{x})}{1-x}$	Ex 10c, pp. 536–537	
5	a. $2\sqrt{5}$ b. $2x$	Ex 6b, c, p. 534		10	$-8-4\sqrt{5}$		Sec 8.4, 11/11

☐   **Next, insert your homework.** Make sure you attempt all exercises asked of you and show all work, as in the exercises above. Check your answers if possible. Clearly mark any exercises you were unable to correctly complete so that you may ask questions later. DO NOT ERASE YOUR INCORRECT WORK. THIS IS HOW WE UNDERSTAND AND EXPLAIN TO YOU YOUR ERRORS.

Section 8.5 Solving Equations Containing Radicals

Before Class:

☐   Read the objectives on page 541.

☐   Read the **Helpful Hint** boxes on pages 541, 543, and 544.

☐   Complete the exercises:

1.   Squaring both sides of an equation can result in an equation that has an

_____ solution that isn't a solution of the original equation.

2.   Read the Solving a Radical Equation Containing Square Roots box on page 542. What is the first step?

During Class:

☐   **Write your class notes.** Neatly write down **all** examples shown as well as key terms or phrases with definitions. If not applicable or if you were absent, watch the Lecture Series (DVD) for this section and do the same (write down the examples shown as well as key terms or phrases). Insert more paper as needed.

Class Notes/Examples	Your Notes

Answers: **1)** extraneous   **2)** Arrange terms so that one radical is by itself on one side of the equation.

Section 8.5 Solving Equations Containing Radicals

Class Notes (continued)	**Your Notes**

(Insert additional paper as needed.)

Copyright © 2013 Pearson Education, Inc.

Section 8.5 Solving Equations Containing Radicals

Practice:

☐   Complete the Vocabulary, Readiness & Video Check on page 544.

☐   Next, complete any incomplete exercises below. Check and correct your work using the answers and references at the end of this section.

Review this example:	**Your turn:**
1. Solve: $\sqrt{x+3} = 5$	**2.** Solve: $\sqrt{x+5} = 2$

To solve this radical equation, use the squaring property of equality and square both sides of the equation.

$$\sqrt{x+3} = 5$$
$$\left(\sqrt{x+3}\right)^2 = 5^2$$
$$x+3 = 25$$
$$x = 22$$

Check: Replace x with 22 in the original equation.

$$\sqrt{x+3} = 5$$
$$\sqrt{22+3} \overset{?}{=} 5$$
$$\sqrt{25} \overset{?}{=} 5$$

True $5 = 5$

Since a true statement results, (22) is the solution.

Review this example:	**Your turn:**
3. Solve: $\sqrt{x} + 6 = 4$	**4.** Solve: $3\sqrt{x} + 5 = 2$

Set the radical by itself on one side of the equation.

$$\sqrt{x} + 6 = 4$$
$$\sqrt{x} = -2$$

$\sqrt{x}$ is the nonnegative square root of x so $\sqrt{x}$ cannot equal -2, and thus the equation has no solution. The same result would be obtained if we continued by applying the squaring property.

$$\sqrt{x} = -2 \qquad \text{Check:} \qquad \sqrt{x} + 6 = 4$$
$$\left(\sqrt{x}\right)^2 = (-2)^2 \qquad\qquad \sqrt{4} + 6 \overset{?}{=} 4$$
$$x = 4 \qquad\qquad \text{False } 2+6 = 4$$

Since 4 *does not satisfy* the original equation, this equation has (no solution.)

Section 8.5 Solving Equations Containing Radicals

Review this example:

5. Solve: $\sqrt{x+3} - x = -3$

Isolate the radical by adding x to both sides. Then square both sides.

$$\sqrt{x+3} - x = -3$$
$$\sqrt{x+3} = x - 3$$
$$\left(\sqrt{x+3}\right)^2 = (x-3)^2$$
$$x+3 = x^2 - 6x + 9$$
$$0 = x^2 - 7x + 6$$
$$0 = (x-6)(x-1)$$
$$0 = x - 6 \text{ or } 0 = x - 1$$
$$6 = x \qquad\qquad 1 = x$$

Check: Replace x with 6 and then 1 in the original equation.

Let $x = 6$.

$$\sqrt{x+3} - x = -3$$
$$\sqrt{6+3} - 6 \overset{?}{=} -3$$
$$\sqrt{9} - 6 \overset{?}{=} -3$$
$$3 - 6 \overset{?}{=} -3$$
$$\text{True} \quad -3 = -3$$

Let $x = 1$.

$$\sqrt{x+3} - x = -3$$
$$\sqrt{1+3} - 1 \overset{?}{=} -3$$
$$\sqrt{4} - 1 \overset{?}{=} -3$$
$$2 - 1 \overset{?}{=} -3$$
$$\text{False} \quad 1 = -3$$

1 is an extraneous solution. The only solution is 6.

Your turn:

6. Solve: $\sqrt{1-8x} - x = 4$

Section 8.5 Solving Equations Containing Radicals

Review this example:

7. Solve: $\sqrt{x-4} = \sqrt{x} - 2$

$$\sqrt{x-4} = \sqrt{x} - 2$$
$$\left(\sqrt{x-4}\right)^2 = \left(\sqrt{x} - 2\right)^2$$
$$x - 4 = x - 4\sqrt{x} + 4$$
$$-8 = -4\sqrt{x}$$
$$2 = \sqrt{x}$$
$$4 = x$$

Check the proposed solution in the original equation. The solution is $\boxed{4.}$

Your turn:

8. Solve: $\sqrt{x-7} = \sqrt{x} - 1$

	Answer	Text Ref	Video Ref			Answer	Text Ref	Video Ref
1	22	Ex 1, p. 541		5	6	Ex 5, pp. 543–544	
2	−1		Sec 8.5, 1/5	6	−1		Sec 8.5, 4/5
3	no solution	Ex 2, pp. 541–542		7	4	Ex 6, p. 544	
4	no solution		Sec 8.5, 2/5	8	16		Sec 8.5, 5/5

☐   **Next, insert your homework.** Make sure you attempt all exercises asked of you and show all work, as in the exercises above. Check your answers if possible. Clearly mark any exercises you were unable to correctly complete so that you may ask questions later. DO NOT ERASE YOUR INCORRECT WORK. THIS IS HOW WE UNDERSTAND AND EXPLAIN TO YOU YOUR ERRORS.

Section 8.5 Solving Equations Containing Radicals

Section 8.6 Radical Equations and Problem Solving

Before Class:

☐   Read the objectives on page 546.

☐   Complete the exercises:

1.   Write the Pythagorean theorem and identify all variables.

2.   The distance d between two points with coordinates (x_1, y_1) and (x_2, y_2) is given by the

     formula _____ .

During Class:

☐   **Write your class notes.** Neatly write down **all** examples shown as well as key terms or
       phrases with definitions. If not applicable or if you were absent, watch the Lecture Series
       (DVD) for this section and do the same (write down the examples shown as well as key terms
       or phrases). Insert more paper as needed.

Class Notes/Examples	**Your Notes**

Answers: **1)** If a and b are the lengths of the legs of a right triangle and c is the length of the
hypotenuse, then $a^2 + b^2 = c^2$.   **2)** $d = \sqrt{(x_2 - x_1)^2 + (y_2 - y_1)^2}$

Section 8.6 Radical Equations and Problem Solving

Class Notes (continued)	**Your Notes**

(Insert additional paper as needed.)

Section 8.6 Radical Equations and Problem Solving

Practice:

☐   Complete the Vocabulary, Readiness & Video Check on page 549.

☐   Next, complete any incomplete exercises below. Check and correct your work using the answers and references at the end of this section.

Review this example:	**Your turn:**
1. Find the length of the leg of the right triangle shown. Give the exact length and a two-decimal approximation.	**2.** Use the Pythagorean theorem to find the unknown side of the right triangle.

Let $a = 2$ meters and b be the unknown length of the other leg. The hypotenuse is $c = 5$ meters.

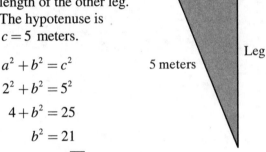

2 meters

5 meters

Leg

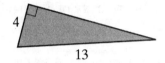

4

13

$a^2 + b^2 = c^2$

$2^2 + b^2 = 5^2$

$4 + b^2 = 25$

$b^2 = 21$

$b = \sqrt{21}$

≈ 4.58 meters

The length of the leg is exactly $\sqrt{21}$ meters or approximately 4.58 meters.

Review this example:	**Your turn:**
3. Find the distance between $(-1,9)$ and $(-3,-5)$.	**4.** Find the distance between $(-3,1)$ and $(5,-2)$.

Use the distance formula with $(x_1, y_1) = (-1,9)$ and $(x_2, y_2) = (-3,-5)$.

$d = \sqrt{(x_2 - x_1)^2 + (y_2 - y_1)^2}$

$= \sqrt{[-3-(-1)]^2 + (-5-9)^2}$

$= \sqrt{(-2)^2 + (-14)^2}$

$= \sqrt{4+196}$

$= \sqrt{200} = 10\sqrt{2}$

The distance is $10\sqrt{2}$ units.

Section 8.6 Radical Equations and Problem Solving

Review this example:

5. A formula used to determine the velocity v, in feet per second, of an object (neglecting air resistance) after it has fallen a certain height is $v = \sqrt{2gh}$, where g is the acceleration due to gravity, and h is the height the object has fallen. On Earth, the acceleration g due to gravity is approximately 32 feet per second per second. Find the velocity of a person after falling 5 feet.

To find the velocity v when $g = 32$ feet per second per second and $h = 5$ feet, use the velocity formula.

$$v = \sqrt{2gh}$$
$$= \sqrt{2 \cdot 32 \cdot 5}$$
$$= \sqrt{320}$$
$$= 8\sqrt{5}$$

The velocity of the person after falling 5 feet is exactly $8\sqrt{5}$ feet per second, or approximately 17.9 feet per second.

Your turn:

6. The formula $v = \sqrt{2.5r}$ can be used to estimate the maximum safe velocity, v, in miles per hour, at which a car can travel if it is driven along a curved road with a radius of curvature, r, in feet. To the nearest whole number, find the maximum safe speed if a cloverleaf exit on an interstate has a radius of curvature of 300 feet.

	Answer	Text Ref	Video Ref		Answer	Text Ref	Video Ref
1	$\sqrt{21} \approx 4.58$ m	Ex 2, p. 546		4	$\sqrt{73}$		Sec 8.6, 3/6
2	$3\sqrt{17}$		Sec 8.6, 2/6	5	$8\sqrt{5}$ feet per second	Ex 5, p. 548	
3	$10\sqrt{2}$	Ex 4, p. 548		6	27 miles per hour		Sec 8.6, 6/6

☐ **Next, insert your homework.** Make sure you attempt all exercises asked of you and show all work, as in the exercises above. Check your answers if possible. Clearly mark any exercises you were unable to correctly complete so that you may ask questions later. DO NOT ERASE YOUR INCORRECT WORK. THIS IS HOW WE UNDERSTAND AND EXPLAIN TO YOU YOUR ERRORS.

Before Class:

☐   Read the objectives on page 552.

☐   Read the **Helpful Hint** box on page 553.

☐   Complete the exercises:

1.   Use the definitions in this section to rewrite each of the following:

   a. $a^{1/n} =$

   b. $a^{m/n} =$

   c. $a^{-m/n} =$

2.   When evaluating $a^{m/n}$, what must be true about a, m, and n?

During Class:

☐   **Write your class notes.** Neatly write down **all** examples shown as well as key terms or phrases with definitions. If not applicable or if you were absent, watch the Lecture Series (DVD) for this section and do the same (write down the examples shown as well as key terms or phrases). Insert more paper as needed.

Class Notes/Examples	**Your Notes**

Answers: **1)** a. $\sqrt[n]{a}$ b. $\sqrt[n]{a^m} = \left(\sqrt[n]{a}\right)^m$ c. $\dfrac{1}{a^{m/n}}$ **2)** a is a positive number, m and n are integers, $n > 0$

Section 8.7 Rational Exponents

Class Notes (continued)	Your Notes

(Insert additional paper as needed.)

Practice:

☐ Complete the Vocabulary, Readiness & Video Check on page 555.

☐ Next, complete any incomplete exercises below. Check and correct your work using the answers and references at the end of this section.

Review this example:	**Your turn:**
1. Write $25^{1/2}$ in radical notation. Then simplify.	**2.** Write $8^{1/3}$ in radical notation. Then simplify.
$25^{1/2} = \sqrt{25} = \boxed{5}$	

Review this example:	**Your turn:**
3. Simplify each expression.	**4.** Simplify each expression.
a. $27^{2/3}$ b. $-16^{3/4}$	a. $32^{2/5}$ b. $64^{3/2}$
a. $27^{2/3} = \left(27^{1/3}\right)^2 = 3^2 = \boxed{9}$ b. The negative sign is *not* affected by the exponent. $-16^{3/4} = \left(-16^{1/4}\right)^3 = -\left(\sqrt[4]{16}\right)^3 = -(2)^3 = \boxed{-8}$	

Review this example:	**Your turn:**
5. Write each expression with a positive exponent and then simplify.	**6.** Simplify each expression.
a. $16^{-3/4}$ b. $32^{-4/5}$	a. $625^{-3/4}$ b. $-16^{-1/4}$
a. $16^{-3/4} = \dfrac{1}{16^{3/4}} = \dfrac{1}{\left(\sqrt[4]{16}\right)^3} = \dfrac{1}{2^3} = \boxed{\dfrac{1}{8}}$ b. $32^{-4/5} = \dfrac{1}{32^{4/5}} = \dfrac{1}{\left(\sqrt[5]{32}\right)^4} = \dfrac{1}{2^4} = \boxed{\dfrac{1}{16}}$	

Section 8.7 Rational Exponents

Review this example:

7. Simplify each expression .Write results with positive exponents only. Assume that all variables represent positive numbers.

   a. $3^{1/2} \cdot 3^{3/2}$ b. $\dfrac{5^{1/3}}{5^{2/3}}$ c. $\left(x^{1/4}\right)^{12}$

Your turn:

8. Simplify each expression using exponential rules. Write results with positive exponents only. Assume that all variables represent positive numbers.

   a. $3^{4/3} \cdot 3^{2/3}$ b. $\dfrac{3^{-3/5}}{3^{2/5}}$ c. $\left(x^{2/3}\right)^{9}$

a. $3^{1/2} \cdot 3^{3/2} = 3^{(1/2)+(3/2)} = 3^{4/2} = 3^2 = \boxed{9}$

b. $\dfrac{5^{1/3}}{5^{2/3}} = 5^{(1/3)-(2/3)} = 5^{-1/3} = \boxed{\left(\dfrac{1}{5^{1/3}}\right)}$

c. $\left(x^{1/4}\right)^{12} = x^{(1/4)12} = \boxed{x^3}$

	Answer	Text Ref	Video Ref		Answer	Text Ref	Video Ref
1	5	Ex 1a, p. 552		5	a. $\dfrac{1}{8}$ b. $\dfrac{1}{16}$	Ex 3b, d, p. 554	
2	2	Sec 8.7, 2/13		6	a. $\dfrac{1}{125}$ b. $-\dfrac{1}{2}$		Sec 8.7, 9–10/13
3	a. 9 b. −8	Ex 2b, c, p. 553		7	a. 9 b. $\dfrac{1}{5^{1/3}}$ c. x^3	Ex 4a, b, c, p. 554	
4	a. 4 b. 512	Sec 8.7, 6–7/13		8	a. 9 b. $\dfrac{1}{3}$ c. x^6		Sec 8.7, 11/13 13/13, 12/13

☐   **Next, insert your homework.** Make sure you attempt all exercises asked of you and show all work, as in the exercises above. Check your answers if possible. Clearly mark any exercises you were unable to correctly complete so that you may ask questions later. DO NOT ERASE YOUR INCORRECT WORK. THIS IS HOW WE UNDERSTAND AND EXPLAIN TO YOU YOUR ERRORS.

Preparing for the Chapter 8 Test

Start preparing for your Chapter 8 Test as soon as possible. Pay careful attention to any instructor discussion about this test, especially discussion on what sections you will be responsible for, etc.

☐   Work the Chapter 8 Vocabulary Check on page 556.

☐   Read your Class Notes/Examples for each section covered on your Chapter 8 Test. Look for any unresolved questions you may have.

☐   Complete as many of the Chapter 8 Review exercises as possible (page 559). Remember, the odd answers are in the back of your text.

☐   **Most important:** Place yourself in "test" conditions (see below) and work the Chapter 8 Test (page 561) as a practice test the day before your actual test. To honestly assess how you are doing, try the following:
- Work on a few blank sheets of paper.
- Give yourself the same amount of time you will be given for your actual test.
- Complete this Chapter 8 Practice Test without using your notes or your text.
- If you have any time left after completing this practice test, check your work and try to find any errors on your own.
- Once done, use the back of your book to check ALL answers.
- Try to correct any errors on your own.
- Use the Chapter Test Prep Video (CTPV) to correct any errors you were unable to correct on your own. You can find these videos in the Interactive DVD Lecture Series, in MyMathLab, and on YouTube. Search Martin-Gay Beginning Algebra and click "Channels."

I wish you the best of luck….Elayn Martin-Gay

Section 9.1 Solving Quadratic Equations by the Square Root Property

Before Class:

☐   Read the objectives on page 565.

☐   Read the **Helpful Hint** boxes on page 567.

☐   Complete the exercises:

1.   Recall the zero factor property: If the product of two numbers is zero, what must be true about at least one of the numbers?

2.   The square root property states that if $x^2 = a$ for $a \geq 0$, then $x =$ _____ or

$x =$ _____.

During Class:

☐   **Write your class notes.** Neatly write down **all** examples shown as well as key terms or phrases with definitions. If not applicable or if you were absent, watch the Lecture Series (DVD) for this section and do the same (write down the examples shown as well as key terms or phrases). Insert more paper as needed.

Class Notes/Examples	**Your Notes**

Answers: **1)** At least one number is zero.   **2)** $\sqrt{a}$; $-\sqrt{a}$

273

Section 9.1 Solving Quadratic Equations by the Square Root Property

Class Notes (continued)	**Your Notes**

(Insert additional paper as needed.)

Section 9.1 Solving Quadratic Equations by the Square Root Property

Practice:

☐   Complete the Vocabulary, Readiness & Video Check on page 568.

☐   Next, complete any incomplete exercises below. Check and correct your work using the
       answers and references at the end of this section.

Review this example:	**Your turn:**
1. Use the square root property to solve $2x^2 = 7$.	**2.** Use the square root property to solve $2x^2 - 10 = 0$.

Review this example:

1. Use the square root property to solve $2x^2 = 7$.

First solve for x^2 by dividing both sides by 2.

$$2x^2 = 7$$
$$x^2 = \frac{7}{2}$$

Then use the square root property.

$$x = \sqrt{\frac{7}{2}} \quad \text{or} \quad x = -\sqrt{\frac{7}{2}}$$

If the denominators are rationalized, then

$$x = \frac{\sqrt{7} \cdot \sqrt{2}}{\sqrt{2} \cdot \sqrt{2}} \quad \text{or} \quad x = -\frac{\sqrt{7} \cdot \sqrt{2}}{\sqrt{2} \cdot \sqrt{2}}$$

$$x = \frac{\sqrt{14}}{2} \qquad x = -\frac{\sqrt{14}}{2}$$

Check both solutions in the original equation. The

solutions are $\left(\dfrac{\sqrt{14}}{2} \text{ and } -\dfrac{\sqrt{14}}{2}\right)$.

Your turn:

2. Use the square root property to solve $2x^2 - 10 = 0$.

Review this example:

3. Use the square root property to solve $(x-3)^2 = 16$.

Instead of x^2, here is $(x-3)^2$. But the square root property can still be used.

$$(x-3)^2 = 16$$
$$x - 3 = \sqrt{16} \quad \text{or} \quad x - 3 = -\sqrt{16}$$
$$x - 3 = 4 \qquad\qquad x - 3 = -4$$
$$x = 7 \qquad\qquad\quad x = -1$$

Check: $(x-3)^2 = 16 \qquad\qquad (x-3)^2 = 16$

$(7-3)^2 \overset{?}{=} 16 \qquad\qquad (-1-3)^2 \overset{?}{=} 16$

$4^2 \overset{?}{=} 16 \qquad\qquad\qquad (-4)^2 \overset{?}{=} 16$

True $16 = 16 \qquad\qquad$ True $16 = 16$

Both 7 and -1 are solutions.

Your turn:

4. Use the square root property to solve $(p+2)^2 = 10$.

Section 9.1 Solving Quadratic Equations by the Square Root Property

Review this example:

5. Use the square root property to solve $(5x-2)^2 = 10$.

$$(5x-2)^2 = 10$$

$$5x-2 = \sqrt{10} \quad \text{or} \quad 5x-2 = -\sqrt{10}$$

$$5x = 2+\sqrt{10} \qquad 5x = 2-\sqrt{10}$$

$$x = \frac{2+\sqrt{10}}{5} \qquad x = \frac{2-\sqrt{10}}{5}$$

Check both solutions in the original equation.

The solutions are $\left(\dfrac{2\pm\sqrt{10}}{5}\right)$.

Your turn:

6. Use the square root property to solve $(3x-7)^2 = 32$.

	Answer	Text Ref	Video Ref			Answer	Text Ref	Video Ref
1	$\dfrac{\sqrt{14}}{2}, -\dfrac{\sqrt{14}}{2}$	Ex 2, p. 566			4	$-2+\sqrt{10}, -2-\sqrt{10}$		Sec 9.1, 4/6
2	$\sqrt{5}, -\sqrt{5}$		Sec 9.1, 3/6		5	$\dfrac{2\pm\sqrt{10}}{5}$	Ex 6, p. 567	
3	$7, -1$	Ex 3, p. 566			6	$\dfrac{7\pm4\sqrt{2}}{3}$		Sec 9.1, 5/6

☐ **Next, insert your homework.** Make sure you attempt all exercises asked of you and show all work, as in the exercises above. Check your answers if possible. Clearly mark any exercises you were unable to correctly complete so that you may ask questions later. DO NOT ERASE YOUR INCORRECT WORK. THIS IS HOW WE UNDERSTAND AND EXPLAIN TO YOU YOUR ERRORS.

Section 9.2 Solving Quadratic Equations by Completing the Square

Before Class:

☐   Read the objectives on page 570.

☐   Read the **Helpful Hint** boxes on page 571.

☐   Complete the exercises:

1.   The process of adding a number to $x^2 + bx$ to form a perfect square trinomial is called

   _____ on $x^2 + bx$.

2.   To solve an equation of the form $(x + a)^2 = c$, use the _____
   property.

During Class:

☐   **Write your class notes.** Neatly write down **all** examples shown as well as key terms or phrases with definitions. If not applicable or if you were absent, watch the Lecture Series (DVD) for this section and do the same (write down the examples shown as well as key terms or phrases). Insert more paper as needed.

<table>
<tr><th>Class Notes/Examples</th><th>Your Notes</th></tr>
</table>

Answers: **1)** completing the square   **2)** square root

Section 9.2 Solving Quadratic Equations by Completing the Square

Class Notes (continued)	**Your Notes**

(Insert additional paper as needed.)

Copyright © 2013 Pearson Education, Inc.

Section 9.2 Solving Quadratic Equations by Completing the Square

Practice:

☐   Complete the Vocabulary, Readiness & Video Check on page 573–574.

☐   Next, complete any incomplete exercises below. Check and correct your work using the answers and references at the end of this section.

Review this example:

1. Solve $x^2 + 6x + 3 = 0$ by completing the square.

First get the variable terms alone by subtracting 3 from both sides of the equation.

$x^2 + 6x + 3 = 0$

$\quad x^2 + 6x = -3$

Next find half the coefficient of the x-term, then square it. Add this result to both sides of the equation. This will make the left side a perfect square trinomial. The coefficient of x is 6, and half of 6 is 3. So add 3^2 or 9 to both sides.

$x^2 + 6x + 9 = -3 + 9$

$\quad (x+3)^2 = 6$

$\quad\quad x + 3 = \sqrt{6} \quad\quad \text{or} \quad x + 3 = -\sqrt{6}$

$\quad\quad\quad x = -3 + \sqrt{6} \quad\quad\quad x = -3 - \sqrt{6}$

Check by substituting $-3 + \sqrt{6}$ and $-3 - \sqrt{6}$ in the original equation.

The solutions are $-3 \pm \sqrt{6}$.

Your turn:

2. Solve by completing the square:

$$x^2 - 2x - 1 = 0$$

Section 9.2 Solving Quadratic Equations by Completing the Square

Review this example:

3. Solve $4x^2 - 8x - 5 = 0$ by completing the square.

$4x^2 - 8x - 5 = 0$

$x^2 - 2x - \dfrac{5}{4} = 0$

$x^2 - 2x = \dfrac{5}{4}$

The coefficient of x is -2. Half of -2 is -1, and $(-1)^2 = 1$. So add 1 to both sides.

$x^2 - 2x + 1 = \dfrac{5}{4} + 1$

$(x - 1)^2 = \dfrac{9}{4}$

$x - 1 = \sqrt{\dfrac{9}{4}} \qquad \text{or} \quad x - 1 = -\sqrt{\dfrac{9}{4}}$

$x = 1 + \dfrac{3}{2} = \dfrac{5}{2} \qquad\qquad x = 1 - \dfrac{3}{2} = -\dfrac{1}{2}$

Both $\dfrac{5}{2}$ and $-\dfrac{1}{2}$ are solutions.

Your turn:

4. Solve by completing the square:

$2y^2 + 8y + 5 = 0$

	Answer	Text Ref	Video Ref		Answer	Text Ref	Video Ref
1	$-3 \pm \sqrt{6}$	Ex 1, p. 571		3	$\dfrac{5}{2}, -\dfrac{1}{2}$	Ex 3, pp. 572	
2	$1 \pm \sqrt{2}$		Sec 9.2, 2/3	4	$-2 \pm \dfrac{\sqrt{6}}{2}$		Sec 9.2, 3/3

☐   **Next, insert your homework.** Make sure you attempt all exercises asked of you and show all work, as in the exercises above. Check your answers if possible. Clearly mark any exercises you were unable to correctly complete so that you may ask questions later. DO NOT ERASE YOUR INCORRECT WORK. THIS IS HOW WE UNDERSTAND AND EXPLAIN TO YOU YOUR ERRORS.

Section 9.3 Solving Quadratic Equations by the Quadratic Formula

Before Class:

☐   Read the objectives on page 575.

☐   Read the **Helpful Hint** boxes on pages 576, 577, and 578.

☐   Complete the exercises:

1.   Write the standard form of a quadratic equation.

2.   Write the quadratic formula using the equation in your answer to exercise 1.

3.   Write the discriminant of the equation in your answer to exercise 1.

During Class:

☐   **Write your class notes.** Neatly write down **all** examples shown as well as key terms or phrases with definitions. If not applicable or if you were absent, watch the Lecture Series (DVD) for this section and do the same (write down the examples shown as well as key terms or phrases). Insert more paper as needed.

Class Notes/Examples	**Your Notes**

Answers: **1)** $ax^2 + bx + c = 0, a \neq 0$ **2)** $x = \dfrac{-b \pm \sqrt{b^2 - 4ac}}{2a}$ **3)** $b^2 - 4ac$

Section 9.3 Solving Quadratic Equations by the Quadratic Formula

Class Notes (continued)	**Your Notes**

(Insert additional paper as needed.)

Section 9.3 Solving Quadratic Equations by the Quadratic Formula

Practice:

☐   Complete the Vocabulary, Readiness & Video Check on page 580.

☐   Next, complete any incomplete exercises below. Check and correct your work using the answers and references at the end of this section.

Review this example:

1. Solve $3x^2 + x - 3 = 0$ using the quadratic formula.

This equation is in standard form with $a = 3$, $b = 1$, and $c = -3$. By the quadratic formula,

$$x = \frac{-b \pm \sqrt{b^2 - 4ac}}{2a}$$

$$x = \frac{-1 \pm \sqrt{1^2 - 4 \cdot 3 \cdot (-3)}}{2 \cdot 3}$$

$$= \frac{-1 \pm \sqrt{1 + 36}}{6}$$

$$= \frac{-1 \pm \sqrt{37}}{6}$$

Check both solutions in the original equation.

The solutions are $\frac{-1 + \sqrt{37}}{6}$ and $\frac{-1 - \sqrt{37}}{6}$.

Review this example:

3. Solve $x^2 = -x - 1$ using the quadratic formula.

First write the equation in standard form.
$$x^2 + x + 1 = 0$$

Next replace a, b, and c in the quadratic formula with $a = 1$, $b = 1$, and $c = 1$.

$$x = \frac{-1 \pm \sqrt{1^2 - 4 \cdot 1 \cdot 1}}{2 \cdot 1}$$

$$= \frac{-1 \pm \sqrt{-3}}{2}$$

There is no real number solution because $\sqrt{-3}$ is not a real number.

Your turn:

2. Solve $3k^2 + 7k + 1 = 0$ using the quadratic formula.

Your turn:

4. Solve $3 - x^2 = 4x$ using the quadratic formula.

Section 9.3 Solving Quadratic Equations by the Quadratic Formula

Review this example:	**Your turn:**
5. Use the discriminant to determine the number of solutions of $3x^2 + x - 3 = 0$. In $3x^2 + x - 3 = 0$, $a = 3$, $b = 1$, and $c = -3$. Then $b^2 - 4ac = (1)^2 - 4(3)(-3) = 1 + 36 = 37$. Since the discriminant is 37, a positive number, this equation has (two distinct real solutions.)	6. Use the discriminant to determine the number of real solutions of the quadratic equation $3x^2 + x + 5 = 0$.

Review this example:	**Your turn:**

7. Use the discriminant to determine the number of solutions of each quadratic equation.
 	
   a. $x^2 - 6x + 9 = 0$ b. $5x^2 + 4 = 0$
 	
a. In $x^2 - 6x + 9 = 0$, $a = 1$, $b = -6$ and $c = 9$.
   $b^2 - 4ac = (-6)^2 - 4(1)(9) = 36 - 36 = 0$
   Since the discriminant is 0, this equation has
   (one real solution.)
 	
b. In $5x^2 + 4 = 0$, $a = 5$, $b = 0$, and $c = 4$.
   $b^2 - 4ac = 0^2 - 4(5)(4) = 0 - 80 = -80$
   Since the discriminant is -80, a negative
   number, this equation has (no real solution.)
	8. Use the discriminant to determine the number of real solutions of each quadratic equation.
	
	   a. $9x^2 + 2x = 0$
	
	   b. $4x^2 + 4x = -1$

	Answer	Text Ref	Video Ref			Answer	Text Ref	Video Ref
1	$\dfrac{-1 + \sqrt{37}}{6}, \dfrac{-1 - \sqrt{37}}{6}$	Ex 1, p. 576			5	2 real solutions	Ex 7, p. 579	
2	$\dfrac{-7 + \sqrt{37}}{6}, \dfrac{-7 - \sqrt{37}}{6}$		Sec 9.3, 1/7		6	no real solution		Sec 9.3, 5/7
3	no real solution	Ex 4, p. 578			7	a. 1 real solution b. no real solution	Ex 8, p. 580	
4	$-2 + \sqrt{7}, -2 - \sqrt{7}$		Sec 9.3, 2/7		8	a. 2 real solutions b. 1 real solution		Sec 9.3, 6–7/7

☐   **Next, insert your homework.** Make sure you attempt all exercises asked of you and show all work, as in the exercises above. Check your answers if possible. Clearly mark any exercises you were unable to correctly complete so that you may ask questions later. DO NOT ERASE YOUR INCORRECT WORK. THIS IS HOW WE UNDERSTAND AND EXPLAIN TO YOU YOUR ERRORS.

Section 9.4 Complex Solutions of Quadratic Equations

Before Class:

☐   Read the objectives on page 585.

☐   Complete the exercises:

1.   The complex number system includes the _____ i.

2.   A complex number can be written in the form _____ .

3.   The complex numbers $a + bi$ and $a - bi$ are called _____ of each other.

During Class:

☐   **Write your class notes.** Neatly write down **all** examples shown as well as key terms or phrases with definitions. If not applicable or if you were absent, watch the Lecture Series (DVD) for this section and do the same (write down the examples shown as well as key terms or phrases). Insert more paper as needed.

Class Notes/Examples	**Your Notes**

Answers: **1)** imaginary unit   **2)** $a + bi$ **3)** complex conjugates

Section 9.4 Complex Solutions of Quadratic Equations

Class Notes (continued)	Your Notes

(Insert additional paper as needed.)

Section 9.4 Complex Solutions of Quadratic Equations

Practice:

☐   Complete the Vocabulary, Readiness & Video Check on page 589.

☐   Next, complete any incomplete exercises below. Check and correct your work using the answers and references at the end of this section.

Review this example:	Your turn:

1. Write each radical as the product of a real number and i.
 	
  a. $\sqrt{-4}$ b. $\sqrt{-20}$
 	
Write each negative radicand as a product of a positive number and -1. Then write $\sqrt{-1}$ as i.
 	
  a. $\sqrt{-4} = \sqrt{-1 \cdot 4} = \sqrt{-1} \cdot \sqrt{4} = i \cdot 2 = \boxed{2i}$
 	
  b. $\sqrt{-20} = \sqrt{-1 \cdot 20} = \sqrt{-1} \cdot \sqrt{20} = i \cdot 2\sqrt{5}$
               $= \boxed{2i\sqrt{5}}$
	2. Simplify each radical. Write it as the product of a real number and i.
	
	  a. $\sqrt{-9}$ b. $\sqrt{-63}$

Review this example:
3. Simplify the sum. Write the result in standard form.
 	
  $(2+3i)+(-6-i)$
 	
Add the real parts and then add the imaginary parts.
 	
$(2+3i)+(-6-i) = \left[2+(-6)\right]+(3i-i)$
                 $= \boxed{-4+2i}$
	Your turn:
	4. Add and simplify. Write the result in standard form.
	
	  $(2-i)+(-5+10i)$

Review this example:
5. Subtract $(11-i)$ from $(1+i)$.
 	
$(1+i)-(11-i) = 1+i-11+i$
             $= (1-11)+(i+i)$
             $= \boxed{-10+2i}$
	Your turn:
	6. Subtract $(2+3i)$ from $(-5+i)$.

Section 9.4 Complex Solutions of Quadratic Equations

Review this example:

7. Find the following products and write in standard form.

   a. $5i(2-i)$ b. $(2+3i)(2-3i)$

a. By the distributive property,

$5i(2-i) = 5i \cdot 2 - 5i \cdot i$

$\qquad\qquad = 10i - 5i^2$

$\qquad\qquad = 10i - 5(-1)$

$\qquad\qquad = 10i + 5 \text{ or } (5+10i)$

b. $(2+3i)(2-3i) = 4 - 6i + 6i - 9i^2$

$\qquad\qquad\qquad = 4 - 9(-1)$

$\qquad\qquad\qquad = (13)$

Your turn:

8. Find the products of the following complex numbers and write in standard form.

   a. $-9i(5i-7)$

   b. $(4-3i)(4+3i)$

Review this example:

9. Write $\dfrac{4+i}{3-4i}$ in standard form.

To write this quotient as a complex number in the standard form $a + bi$, find an equivalent fraction whose denominator is a real number. Multiplying both numerator and denominator by the denominator's conjugate gives a fraction that is an equivalent fraction with a real number denominator.

$\dfrac{4+i}{3-4i} = \dfrac{4+i}{3-4i} \cdot \dfrac{3+4i}{3+4i}$

$\qquad = \dfrac{12+16i+3i+4i^2}{9-16i^2}$

$\qquad = \dfrac{12+19i+4(-1)}{9-16(-1)}$

$\qquad = \dfrac{12+19i-4}{9+16} = \dfrac{8+19i}{25}$

$\qquad = \left(\dfrac{8}{25} + \dfrac{19}{25}i \right)$

Your turn:

10. Find the quotient of the complex expression $\dfrac{7-i}{4-3i}$ and write it in standard form.

Section 9.4 Complex Solutions of Quadratic Equations

Review this example:

11. Solve $(x+2)^2 = -25$.

Begin by applying the square root property.

$(x+2)^2 = -25$

$x+2 = \pm\sqrt{-25}$

$x+2 = \pm 5i$

$x = -2 \pm 5i$

The solutions are $-2+5i$ and $-2-5i$.

Your turn:

12. Solve $(x+1)^2 = -9$.

Review this example:

13. Solve $m^2 = 4m - 5$.

Write the equation in standard form and use the quadratic formula to solve.

$m^2 = 4m - 5$

$m^2 - 4m + 5 = 0$

Apply the quadratic formula with $a = 1$, $b = -4$, and $c = 5$.

$m = \dfrac{4 \pm \sqrt{16 - 4 \cdot 1 \cdot 5}}{2 \cdot 1}$

$= \dfrac{4 \pm \sqrt{-4}}{2}$

$= \dfrac{4 \pm 2i}{2}$

$= \dfrac{2(2 \pm i)}{2} = 2 \pm i$

The solutions are $2 - i$ and $2 + i$.

Your turn:

14. Solve $2m^2 - 4m + 5 = 0$.

289

Section 9.4 Complex Solutions of Quadratic Equations

	Answer	Text Ref	Video Ref		Answer	Text Ref	Video Ref
1	a. $2i$ b. $2i\sqrt{5}$	Ex 1a, c, p. 585		8	a. $45 + 63i$ b. 25		Sec 9.4, 5–6/9
2	a. $3i$ b. $3i\sqrt{7}$		Sec 9.4, 1–2/9	9	$\dfrac{8}{25} + \dfrac{19}{25}i$	Ex 6, p. 587	
3	$-4 + 2i$	Ex 3a, p. 586		10	$\dfrac{31}{25} + \dfrac{17}{25}i$		Sec 9.4, 7/9
4	$-3 + 9i$		Sec 9.4, 3/9	11	$-2 + 5i$, $-2 - 5i$	Ex 7, p. 588	
5	$-10 + 2i$	Ex 4, p. 586		12	$-1 + 3i$, $-1 - 3i$		Sec 9.4, 8/9
6	$-7 - 2i$		Sec 9.4, 4/9	13	$2 - i$, $2 + i$	Ex 8, p. 588	
7	a. $5 + 10i$ b. $13 + 0i$	Ex 5a, c pp. 586–587		14	$\dfrac{2 - i\sqrt{6}}{2}$, $\dfrac{2 + i\sqrt{6}}{2}$		Sec 9.4, 9/9

☐　**Next, insert your homework.** Make sure you attempt all exercises asked of you and show all work, as in the exercises above. Check your answers if possible. Clearly mark any exercises you were unable to correctly complete so that you may ask questions later. DO NOT ERASE YOUR INCORRECT WORK. THIS IS HOW WE UNDERSTAND AND EXPLAIN TO YOU YOUR ERRORS.

Section 9.5 Graphing Quadratic Equations

Before Class:

☐ Read the objectives on page 590.

☐ Read the **Helpful Hint** boxes on pages 591, 592 and 595.

☐ Complete the exercises:

1. The graph of $y = x^2$ is a smooth curve called a _____ .

2. The graph of $y = x^2$ is symmetric about the _____ - axis.

3. For the graph of $y = ax^2 + bx + c$,

   if a is positive, the graph opens _____ ;

   if a is negative, the graph opens _____ .

During Class:

☐ **Write your class notes.** Neatly write down **all** examples shown as well as key terms or phrases with definitions. If not applicable or if you were absent, watch the Lecture Series (DVD) for this section and do the same (write down the examples shown as well as key terms or phrases). Insert more paper as needed.

Class Notes/Examples	Your Notes

Answers: **1)** parabola   **2)** y **3)** upward, downward

Section 9.5 Graphing Quadratic Equations

Class Notes (continued)

Your Notes

(Insert additional paper as needed.)

Practice:

☐   Complete the Vocabulary, Readiness & Video Check on page 596.

☐   Next, complete any incomplete exercises below. Check and correct your work using the answers and references at the end of this section.

Review this example:

1. Graph: $y = -2x^2$

Select x-values and calculate the corresponding y-values. Plot the ordered pairs found. Then draw a smooth curve through those points.

When the coefficient of x^2 is negative, the corresponding parabola opens downward. When a parabola opens downward, the vertex is the highest point of the parabola. The vertex of this parabola is $(0,0)$ and the axis of symmetry is the y-axis.

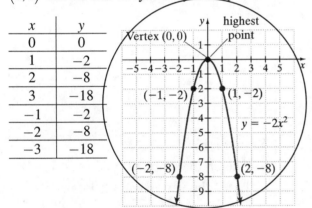

x	y
0	0
1	-2
2	-8
3	-18
-1	-2
-2	-8
-3	-18

Your turn:

2. Graph $y = 2x^2$ by finding and plotting ordered pair solutions.

Review this example:

3. Graph: $y = x^2 - 6x + 8$

In the equation $y = x^2 - 6x + 8$, $a = 1$ and $b = -6$. The x-coordinate of the vertex is

$$x = \frac{-b}{2a} = \frac{-(-6)}{2 \cdot 1} = 3.$$

To find the corresponding y-coordinate, let $x = 3$ in the original equation.

$$y = x^2 - 6x + 8 = 3^2 - 6 \cdot 3 + 8 = -1$$

The vertex is $(3, -1)$ and the parabola opens upward since a is positive.

(solution continued on the next page)

Your turn:

4. Sketch the graph of
$y = 2x^2 - 11x + 5$. Label the vertex and the intercepts.

Section 9.5 Graphing Quadratic Equations

To find the x-intercepts, let $y = 0$.

$0 = x^2 - 6x + 8 = (x-4)(x-2)$

The x-intercepts are $(4,0)$ and $(2,0)$.

Let $x = 0$ in the original equation, then $y = 8$ and the y-intercept is $(0,8)$. Now plot the vertex $(3,-1)$ and the intercepts $(4,0)$, $(2,0)$, and $(0,8)$. Then sketch the parabola.

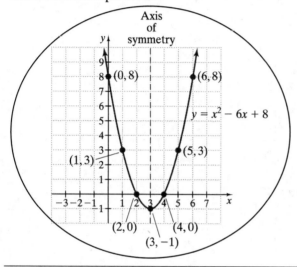

Review this example:

5. Graph: $y = x^2 + 2x - 5$

In the equation $y = x^2 + 2x - 5$, $a = 1$ and $b = 2$. Using the vertex formula, find that the x-coordinate of the vertex is

$$x = \frac{-b}{2a} = \frac{-2}{2 \cdot 1} = -1.$$

The y-coordinate of the vertex is

$$y = (-1)^2 + 2(-1) - 5 = -6$$

Thus, the vertex is $(-1,-6)$.

To find the x-intercepts, let $y = 0$.

$0 = x^2 + 2x - 5$

This cannot be solved by factoring, so use the quadratic formula.

(solution continued on the next page)

Your turn:

6. Sketch the graph of $y = -x^2 + 4x - 3$. Label the vertex and the intercepts.

$$x = \frac{-2 \pm \sqrt{2^2 - 4(1)(-5)}}{2 \cdot 1}$$

$$x = \frac{-2 \pm \sqrt{24}}{2}$$

$$x = \frac{-2 \pm 2\sqrt{6}}{2}$$

$$x = \frac{2\left(-1 \pm \sqrt{6}\right)}{2} = -1 \pm \sqrt{6}$$

The x-intercepts are $\left(-1 + \sqrt{6}, 0\right)$ and $\left(-1 - \sqrt{6}, 0\right)$.

Use a calculator to approximate and easily graph these intercepts.
$-1 + \sqrt{6} \approx 1.4$ and $-1 - \sqrt{6} \approx -3.4$

To find the y-intercept, let $x = 0$ in the original equation to find that $y = -5$. Thus the y-intercept is $(0, -5)$.

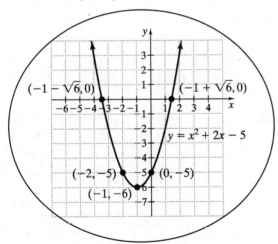

Section 9.5 Graphing Quadratic Equations

	Answer	Text Ref	Video Ref
1		Ex 1, p. 591	
2			Sec 9.4, 1/3
3		Ex 3, pp. 593–594	
4			Sec 9.4, 3/3
5		Ex 4, pp. 594–595	
6			Sec 9.4, 2/3

☐ **Next, insert your homework.** Make sure you attempt all exercises asked of you and show all work, as in the exercises above. Check your answers if possible. Clearly mark any exercises you were unable to correctly complete so that you may ask questions later. DO NOT ERASE YOUR INCORRECT WORK. THIS IS HOW WE UNDERSTAND AND EXPLAIN TO YOU YOUR ERRORS.

Preparing for the Chapter 9 Test

Start preparing for your Chapter 9 Test as soon as possible. Pay careful attention to any instructor discussion about this test, especially discussion on what sections you will be responsible for, etc.

☐   Work the Chapter 9 Vocabulary Check on page 597.

☐   Read your Class Notes/Examples for each section covered on your Chapter 9 Test. Look for any unresolved questions you may have.

☐   Complete as many of the Chapter 9 Review exercises as possible (page 601). Remember, the odd answers are in the back of your text.

☐   **Most important:** Place yourself in "test" conditions (see below) and work the Chapter 9 Test (page 602) as a practice test the day before your actual test. To honestly assess how you are doing, try the following:

- Work on a few blank sheets of paper.
- Give yourself the same amount of time you will be given for your actual test.
- Complete this Chapter 9 Practice Test without using your notes or your text.
- If you have any time left after completing this practice test, check your work and try to find any errors on your own.
- Once done, use the back of your book to check ALL answers.
- Try to correct any errors on your own.
- Use the Chapter Test Prep Video (CTPV) to correct any errors you were unable to correct on your own. You can find these videos in the Interactive DVD Lecture Series, in MyMathLab, and on YouTube. Search Martin-Gay Beginning Algebra and click "Channels."

I wish you the best of luck….Elayn Martin-Gay

The Student Organizer guides students through three important parts of studying effectively—note-taking, practice, and homework.

It is designed to help you organize your learning materials and develop the study habits you need to be successful. The Student Organizer includes:

- How to prepare for class
- Space to take class-notes
- Step-by-step worked examples
- Your Turn exercises (modeled after the examples)
- References to the Martin-Gay text and videos for review
- Helpful hints and directions for completing homework assignments

A flexible design allows instructors to assign any or all parts of the Student Organizer.

The Student Organizer is available in a loose-leaf notebook-ready format. It is also available for download in MyMathLab.

For more information, please go to

www.pearsonhighered.com

www.mypearsonstore.com
(search Martin Gay, Beginning Algebra, Sixth Edition)

ISBN-13: 978-0-321-78521-3
ISBN-10: 0-321-78521-5

PEARSON   ALWAYS LEARNING